Cryptids

The Legendary Stories of the Loch Ness Monster From an Unbiased View

(Cryptids and Other Tales of the Dark)

Lauren Bryant

Published By **Oliver Leish**

Lauren Bryant

Cryptids: The Legendary Stories of the Loch Ness Monster From an Unbiased View(Cryptids and Other Tales of the Dark)

ISBN   978-1-998901-03-6

Legal & Disclaimer

The information contained in this ebook is not designed to replace or take the place of any form of medicine or professional medical advice. The information in this ebook has been provided for educational & entertainment purposes only.

The information contained in this book has been compiled from sources deemed reliable, and it is accurate to the best of the Author's knowledge; however, the Author cannot guarantee its accuracy and validity and cannot be held liable for any errors or omissions. Changes are periodically made to this book. You must consult your doctor or get professional medical advice before using any of the suggested remedies, techniques, or information in this book.

Table of contents

Chapter 1: Mythical Creatures

Many legends can be traced back through literary works and mythologies. These mythical creatures have become popular due to the spreading of literature and storytelling. Below are five of the most popular mythological creatures.

The Phoenix

The mythological bird known as the phoenix can be found in Indian, Egyptian and Greek mythology. According to legend, phoenix's innate ability to know when its 1,000-year lifespan is ending means that it can anticipate the end of it. The bird will make a funeral for itself using cinnamon or another aromatic material, light it on fire and allow itself to be consumed. The ashes are then resurrected by a new phoenix, who rises to start a new chapter in his life. Most mythology associates the risingphoenix with the rising sun.

Centaur

The centaur is an equine hybrid that combines a horse with a human. Although the centaur has a human-like upper body, its legs and lower body look more like those of horses. The mythology that inspired the centaur comes from Greek mythology. It has been enthralled people for many generations due to its numerous stories. Due to their hybrid nature of man-wild animal, the centaurs are often described as raucous and prone to fighting and binging. This can lead them into problems with people. One exception to this typecasting, according to Greek mythology. A centaur named Chiron was portrayed as a wise and well-respected healer.

Mermaids

Mermaids, long-haired fish maidens, are known for their seductive charm and seducing abilities. Since ancient times (at least 3,000 BC), sailors have been able to see these fascinating animals. Their head and trunk are that of a young girl maiden, while their lower body is that of a fish. While some accounts recount good beings who rescue individuals from drowning, others speak of evil creatures that lure men into their peril by singing beautiful songs that cause them to crash against the rocks and die.

Leviathan

Leviathan, a name that refers to an angry sea creature which is mentioned multiple times in Old Testament Scriptures, is Leviathan. It has been described by some as a large sea snake with scaly and glowing skin. Others describe it as a large, muscular crocodile. According to some, God created the male and female Leviathans but destroyed their female counterparts to prevent the oceans being overrun by sea monsters. Leviathan is a mythological beast that has been known to destroy ships, kill and consume many people.

Dragons

The dragon story dates back at most 4,000 years. The dragon is a giant, flying creature that breathes fire and has poisonous nostrils. The dragon is often depicted by Western mythology as a dragon that kidnaps fair girls. This spurs a young knight to get his guns and fight the monster to save the girl. On the contrary, dragons in China are loved and respected as loving and intelligent creatures that defend humankind. They are often regarded as symbols for bravery.

Seven Cryptids Found in The American

The term "cryptid", a unproven organism, is what the term refers too. Even though the word "cryptid" became very popular in the 1980s, supernatural sightings and reports have been plagued the world for years. This ROTW highlights some of North America's most well-known cryptids. They can be found in forests and rivers across North America, as well as on Bureau of Land Management land.

The Midwest is filled of intriguing and mysterious things. The Midwest is filled with dense forest, deep lake, and ample open space. It's hard to know what lies beyond. These cryptids hide in the shadows of our everyday lives as we travel to work, school, and social events.

Cryptids' behavior and environment can make a difference in different parts of the country. Cryptids which live in water are more likely to have skin similar to an amphibian reptile. Many people who live in the Midwest or other cold regions of the world have thick hair, and are resilient to the elements. We've been roaming the Midwest searching for new spooky spots and elusive cryptoids. The following eight cryptids topped our list as the most dangerous and well-known in Midwest.

BEAST OF BRAY ROOAD (WISCONSIN), HUMANOID & BEAR/WOLF LIKE CHARACTERISTICS.

Bigfoot is described as the Beast of Bray Road. According to those who have witnessed it, it is bipedal. It resembles a cross between an animal and a beast. A disturbing tale involves a woman who came to a stop on a dark country road and collided with what she thought was an animal. After exiting her van to investigate further, she discovered that the large wolf-like monster had jumped on its hind legs towards her and was now charging toward her. She ran back to her car panicking and pressed down the accelerator when the beast leapt onto its trunk. The beast, unable and unwilling to hold on, began to slide off the car as she drove.

A HUMANOID FROG FROGMAN - LOVELAND, OHIO

The Loveland Frogman displays all the characteristics and behaviors of a frog. He stands on two-legged legs and stalks Ohio rivers' banks. The beast is described as having leathery, webbed skin, feet, and an froglike head. Two police officers originally saw it, and both remember the beast jumping over the railings like a man.

MONSTER OF TUTTLE OTTOMS, ILLINOIS: AN APE - LIKE ANTEATER

The Tuttle Bottoms Monster of Illinois is an intriguing cryptid you can study. It is thought to live in Illinois's wetlands. The monster has a long snout and resembles an anteater. Although there is not much information about the cryptid yet, at least 50 sightings of it have been reported in and around Tuttle Bottoms. One resident stated that a U.S. Department of Agriculture employee informed him they were investigating this monster. Sightings have diminished, but there's no doubt something is lurking at Tuttle Bottom, Illinois.

INDIANA'S GREENCLAWED GHOST: UNDERWATER CREATIVE

The Green Clawed Beast of Indiana is a water beast that has been similar to the Creature in the Black Lagoon. Is it possible that the Green Clawed Beast was inspired by the Creaturefrom the Black Lagoon or that the Gillman photograph for the Creaturefrom the Black Lagoon was influenced by the Green Clawed Beast? The Creature from the Black Lagoon was only seen once, when a woman and a friend were scuba diving in the Ohio River. The female was pulled beneath the water surface by an unrevealed but powerful grip. Only her buddy was able to free her. The woman was again dragged underneath the surface, before she was finally freed. Reports state that a man, purporting to represent the Air Force, visited their homes shortly after the incident and advised them not again to talk about it. Although the Green Clawed Beast can only be sighted once in Ohio, it has caused fear among people who swim in the river.

MAN-LIKE EXTRATERRESTRIALS IN THE HOPKINSVILLE GOBLINS (KENTUCKY)

In 1955, twelve men ran to Hopkinsville, Kentucky police station in an attempt to warn the public about an alien invasion. The group claims that they witnessed a flying Saucer land nearby their residence. They claimed that small, monkey-like males approached the home in an attempt to gain entry. Also described were large, brightly lit eyes, pointed ears, and webbed arms. They claimed that they had fought them off with firearms. However, the location of the altercation revealed very little beyond a few bullets wasted.

HODAG: HODAG is a WISCONSIN ELEPHANT/DINOSAUR BEAST WHICH HAS A FROG-FACED FUNCTION

The Hodag is fascinating because of its appearance. The creature is well-known throughout Wisconsin. It features a frog's head, an Elephant's face and thick, short legs. It also has claws, claws, the back like a dinosaur and huge teeth. It is unclear what happened to Hodag. However, Eugene Shepard was a legendary land surveyor who allegedly organized a group of people to kill the beast with Dynamite. Many different stories exist about the Hodag. However, some claim Shepard invented the entire story. In either case the Hodag had a permanent presence established in Rhinelander.

MONSTER OF POPE LICK (KENTUCKY): HALF MAN, HALF GOAT, HALF SHEEP

The Pope Lick Monster of Kentucky, which is a cross between man, goat and sheep, lives underneath a railroad bridge at Pope Lick Creek. These are some of the most horrific reports on the Pope Lick Monster. According to reports this beast can lead to manic episodes and make people jump from the bridge. Others believe that the monster uses the power of hypnosis and hypnosis, to lure victims onto the bridge. There have been many fatalities from accidents at this location.

After you have learned the basics, are you ready to hunt? We're on a global tour looking for the most cryptids, both famous and unknown. We expect to have an experience similar as those described in previous stories. We invite you to join us at an event, where you can discover the truth that few people dare to admit. You should be cautious!

Carpeting of Certain Species

Cryptozoologists examine cryptids. One of the many monsters Cryptozoologists have investigated include Bigfoot (Nessie the Loch Ness Monster), El Chupacabra Mothman (Mothman), El Chupacabra and Nessie El Loch Ness Monster). But did you know many of today's modern-day creatures were once classified cryptids as well?

1. Komodo dragon. Until 1910, every serious scientist would have laughed off stories about a gigantic reptile that lived on the Indonesian Island of Komodo. After Lieutenant Steynvansbroek arrested and killed one, all that changed was for the better. W. Douglas Burden - an explorer - was not satisfied with a dead specimen. So he went to the Island in search of a live one. He returns home with a few preserved specimens as well as two Komodo dragons. The Bronx Zoo was home to the dragons. Merian C.

Cooper was moved by their presence and inspired to write 1933's King Kong.

2. Platypus (if you're new to the platypus and have a photograph of it, it's not difficult to see that it's a mix of an otter and a duck. Naturalists, scientists, and most Europeans in 18th century Europe were skeptical that such an animal existed. Captain John Hunter (second governor of New South Wales) sent a platypus sketch and pelt to Europeans in 1798 shortly after it was first discovered. Robert Knox, a doctor, anatomist. ethologist. and physician was certain the platypus pelt was forged by an Asian taxidermist. George Shaw, a botanist who believed the platypus could be real, encouraged him to cut into the fur in search of stitches. After many years of exploration, the platypus' existence was confirmed.

3. Okapi-sometimes known as the forest or giraffe okapi, the okapi is a mix of a doe, a donkey a deer, and a horse. Its closest genetic relative is the Giraffe.

Europeans nicknamed this creature "African Unicorn", in the 18th or 19th centuries. Since the okapis can be found in central Africa, they were well-known by Africans. Opakis are cryptid because they are difficult to find and rarely seen. Sir Harry Johnston, a British Museum cryptid, discovered an okapi bone and skin.

4. Gorilla: You didn't expect to see one on this list. European explorers considered gorillas monster-like. Hanno, who was a Greek adventurer and explorer, saw the first gorilla seen by a non African in the fifth-century BC. Today, scientists think Hanno was referring to chimps and baboons. His translators called the animals he saw "gorillas" (strange indeed).).

Another explorer, Andrew Battel, reported seeing "monsters", similar to humans, come to his campfire each morning when he returned from work. However, he was forced to emphasize the fact that they could not add more wood or fuel to the fire to ensure it continued to burn. Gorillas were cryptids up until Thomas Savage's 1847 discovery of gorilla bones from Libera. He also co-authored a formal account of the new species along with Jeffries Wyman, an Harvard anatomist. They named it Gorillagorilla. Paul du Chaillu, an Anthropologist, was able to hunt living gorillas in the ten years that followed for specimens. Up until 1902, when Robert von Berigne (German captain) identified the first mountain gorilla specimen, it remained a cryptid.

5. Gigantic giant squid -- Many people still believe that the giant-sized squid can be classified as a cryptozoological animal. Giant squid live in deep ocean habitats that aren't accessible to humans, as with many cryptids. Japan's National Science Museum was the first to capture photographs of a large squid. The 24-foot female giant shrimp was captured in Japan by researchers in 2004. Every few weeks, another news article describes

the sighting of a dead giant squid. While many people believe the huge squid to be fake, scientific data suggests otherwise.

Bondegezou -- Bondegezou (or Bondegezou) is one the Moni people's ancestor spirit in Western Papua New Guinea. Because of its ties into Western Papua New Guinea mythology and bondegezou, the bondegezou remained a cryptid throughout history. Australian scientist Tim Flannery photographed the bondegezou's first photograph in the 1980s. Flannery described his creature as a tree dwelling marsupial with the appearance a small-sized man. It is covered in a black and brown fur and even walks on two wheels! Unfortunately, the bondegezou was listed as an endangered species.

7. Kangaroo -- Kangaroos can be difficult to believe were originally classified as cryptoids. Amerigovespucci was the first to describe a kangaroo during his journey along Australia's southern coast in 1499.

He described it to be a horrifying creature with a head of a fox, hands of man, and a tail of a monkey. It also had a bag for its babies. Francisco Pelsaert had captured a kangaroo on 1629. But it perished in the voyage. The kangaroo never became a truly endangered species until Sir Joseph Banks found it on Captain Cook's voyage in 1770.

This section will focus on a few cryptids, which have been made official species. Cryptozoology is a way to see that there are many animals on the planet that we do not understand. But just because something is not understood does not mean it cannot exist. Earth is a fascinating planet, filled with scary cryptids and other strange creatures that lurk in places we might not expect. There is still hope of more cryptids being discovered, including Bigfoot (The Jersey Devil), El Chupacabra and The Orange Pendek.

Cryptids With Potential Existence

Cryptozoology can be defined as an organism that science has yet to prove or disprove. These creatures, collectively called "cryptids", include the Loch Ness Monsters, Bigfoot, and the Himalayan Yeti. However they are not the only ones. Nearly every country or region of the planet has a mythical creature or monster, from giant bats found in Java to large water hounds found in Ireland.

1. AHOOL

Ahools is a giant predatory bat that lives in the Indonesian jungles of Java. Ahools reach more than ten feet in length and can be seen from the ground. Although many people mistakenly believe that ahools were misidentified as eagles, owls and other great birds-of-prey that share the same jungles, some sources state that they are real and may even be an isolated species, possibly descended from pterosaurs.

2. AKKOROKAMUI

Japan's Ainu people have believed for years that Volcano Bay, which is off the southern coast Hokkaido, houses a huge octopus called the Akkorokamui. Many sightings and reports of the creature have been made over the years. John Batchelor, a British missionary who worked in Hokkaido in 1900s, describes one instance in his book The Ainu and Their Folklore. The monster attacked three fishermen and their boat. The men ran in dismay because of the horrendous smell, and not fear. Whatever the case, they were so afraid they refused to get up in the morning and ate their breakfast, and they laid there shaking and pale.

3. ALTAMAHA-HA

The Altamahaha (or Altamaha) is a river monster of 20-30 feet that lives near Darien, Georgia. It has enormous flippers and a seal like snout. Although numerous sightings reported of the Altamahaha have been reported over time, the fact Darien was founded in 1736 by a group Scottish Highlanders as New Inverness suggests that the mythology of the Loch

Ness Monster legends is most likely a descendant.

4. DOBHAR-CHU

The Dobharch, or "water hound", is a mythical, otter-like creature found in isolated freshwater bodies and rivers throughout Ireland. The Dobharch, also known as the "water dog", is a mixed-dog-fish hybrid. It has a long and thick furry body that can move in and out of water quickly. According to one myth, it can even keep up a galloping horses. The sightings of the monster go back many centuries. There are at minimum two gravestones in Ireland, one dating back to 1722 (County Leitrim), that record victims of attacks and deaths by Dobharchs.

5. EMELA-NTOUKA

Many Central African tribal people believe that the Congo basin's marshes harbor a huge semi-aquatic beast known as the emela ntouka. The emela—ntouka, which looks like, but is bigger than, a Hippopotamus, is equipped with a single long bony, bony tusk, or horn in the middle of its forehead. Although it appears to be herbivorous, it has a reputation as being dangerously aggressive when disturbed. It has been known to turn onto and kill other animals larger than itself, similar to the hippopotamus.

6. FILIKO TERAS

According to legend, the sea monster To Filiko Teras (or "the friendly creature") lives in waters near Cyprus's Cape Greco National Park. Although the monster's name implies it has never attacked humans it has earned a reputation for being a nuisance to fishermen and turning boats smaller than it. The Filiko Teras' stories are most likely inspired from the Greek legend of Scylla. It is a monstrous ocean monster that attacked Odysseus's vessel in The Odyssey. However, sightings may be mistakenly mistaken as squids and octopuses.

7. GROOTSLANG

The grootslang or "big Snake" is a legendary creature said to have lived in the Richtersveld, an area of mountainous desert in northwest South Africa. In indigenous folklore the grootslangs consisted of an animal that looked like an elephant's head, and front, and a serpent's back and tail. While the grootslangs seemed to have vanished when the Earth was formed in the year 2000, legends show that some survived the event and sought refuge among the Northern Cape's deepest underground caves. Legends of huge, tusked snakes have been told in South African folklore. Stories about real-life sightings and disappearances of British diamond magnates in the Richtersveld caverns in 1917 are often attributed as a result of a grootslang.

8. JERSEY DEVIL

Jersey Devil is the cryptid known to inhabit New Jersey's Pine Barrens. According to tradition the creature was the thirteenth unwelcome child of

Mother Leeds, the original state immigrant. Because she and her husband could not have another child, she offered her son to him shortly after his birth in 1735. In the Pine Barrens, there have been hundreds upon hundreds of sightings over the years of this grotesque, two-legged, hooved monster. This included one incident in the winter 1909 where a trail of hooved footprints mysteriously appeared under fences, and even on rooftops.

9. MAPINGUARI

The Mapinguari is a gigantic ape-like creature that lives in the rainforests bordering Brazil and Bolivia. The Mapinguari can reach 8 feet high and has a dense red fur covering on its belly and head. It also has long, curved claws. Legends claim that it has a second, larger mouth at the stomach. Mapinguaris are said to be able to raise their hind legs as bears to protect themselves from potential hunters and release a foul-smelling smell when people approach them. The New York Times recently reported that a sighting was made in 2007.

10. OGOPOGO

The Ogopogo was a huge water serpent thought to be found in British Columbia's Lake Okanagan. Ogopogo sightings go back to the 1800s. The monster was first given the Native American name n'haitaka, which translates roughly as "water demon". Ogopogo did not become popular until 1920s. It was named after the title of The Ogo-Pogo. The Funny Foxtrot. His mother was a earwig, his dad a whale, / so I'm going sprinkle some salt on his tail.

11. OLGOI-KHORKHOI

Mongolian olgoikhorkhoi means "big intestine worm," though this underground cryptid is four feet long and more like a giant Earthworm than a parasitic Tapworm. The "Mongolian dungworm" is another name for the olgoikhorkhoi. This worm appears to live below the sands of southern Gobi Desert.

It only emerges during warm summer months or when ground is too moist for survival. Native Mongolians claim they have seen the worms over the centuries. They say that the olgoi khorkhoi has the ability to spit venom, or even acid, and that its body is coated with toxic slime which causes it to die quickly.

12. MOMO

Momo, which is shorthand for Missouri Monster, is a cryptic giant apeman who is believed to haunt the Missouri River. Momo was first sighted in 1971. It is described as being between 7-8 feet tall with a large pumpkin-shaped face and fully covered in dark fur, from head to feet. According to some stories, it can be extremely hostile. It can, like the South American Mapinguari's, emit an unpleasant smell to deter attackers.

13. SHUCK

Mythology of the British Isles is filled with stories about mysterious black dogs who haunt villages and towns. The Shuck-a monstrously black hound claimed to reside in East Anglia, on England's far eastern coast--is perhaps the most well-known. It allegedly attacked the church at Bungay in Suffolk in 1577 during a severe thunderstorm. The Shuck, a large, black dog, entered the church door to rescue the villagers from the storm. He killed a father, his son, and tore apart one of the support pillars for the church steeple. As it fled, Shuck apparently left burnt traces in the church door's wood. These can still being seen today.

14. TATZELWURM

Tatzelwurms - lizard-like creatures - are believed to be in the Alps' most remote parts. They are between two and five-foot tall with a large head and gaping mouth. Their forelimbs tend to be shorter and equipped with strong claws. However, their hind legs have long snake-like talons. They are known to be tatzelwurms and arassas from Germany, stollenwurms and stollenwurms respectively in France and Italy.

However, many sightings of them have been documented throughout the Alps.

15. TESSIE

Tahoe Tessie is a lake beast that is believed to reside in Lake Tahoe. Tessie sightings can be traced back to 1913 and usually describe a massive, snake-like monster swimming at speeds fast enough to keep pace with sailboats. Surprisingly Tessie sightings in odd-numbered numbered years are more frequent than in even.

16. YOWIE

Yowies are a kind of ape like Bigfoot. They are believed to live in Australia's Outback. They are usually tall and stocky with thick black fur or dark red fur covering their bodies from head to foot. However, most yowie sightings seem to indicate that the animals can be timid and easily startled. They can also be seen yelling bloodcurdling screams at people, according to some accounts. They are believed to be mythical today. Yet, sightings of these creatures were so common in nineteenth-century Australia that Herbert J. McCooey, an Australian scholar and amateur adventurer, wrote to Sydney's Australian Museum, offering to capture it for US$40 (approximately US$3000/PS1,800). He fell short.

Famous Cryptids From The United States

Cryptids (such as Sasquatch) are creatures whose existence was affirmed but never confirmed. Many of the most famous cryptids were born out of folklore. These stories were passed down to their descendants at a local level. Contrary the common misconception, cryptids can be either magical, legendary or unusual. Many popular animals have these traits, however.

These are the five most frequent cryptids encountered in the United States.

Bigfoot

Bigfoot is also known as Sasquatch. This bipedal ape like monster stands about 6-9ft tall and has hair that can either be black, dark brown or reddish. The creatures' enormous footprints can reach up to 24 inches long and 8 inches wide.

Sightings

Patterson-Gimlin's film is well-known for a Bigfoot sighting. This is an American short film about an unidentified subject. It is claimed that the filmmaker is a Bigfoot. The footage was filmed in Northern California in 1967. There have been many attempts to authenticate it or disprove.

Roger Patterson and Robert Gimlin helmed the film. Patterson, who had died from cancer in 1972 was "consistently maintained throughout that the movie the truth was true." Gimlin repeatedly denied any involvement in a hoax involving Patterson. He largely avoided public discussion on the matter until 2005, after which he resumed giving interviews to Bigfoot conferences and giving interviews.

According to their accounts Patterson, Gimlin, were riding northeast on horseback down the east bank Bluff Creek on Friday 20th October 1967. They "came upon an upturned tree that had a massive root network nearly as high as a bed" at a bend on the creek. As they rounded the bend, they noticed a man standing near the creek to their left. Gimlin later stated he was in minor shock when he first witnessed the apparition.

Many other sightings followed the initial sightings. Most were determined to be hoaxes.

Goatman

Urban legend claims that the Goatman (a half-goat, part-man creature) is a mixture of a goat's head and hindquarters as well as a human body.

The Goatman's story begins at the Beltsville Agricultural Research Center with a scientist who conducted experiments on goats. However, one of these failed and he was mutated into a goatlike creature. He began to attack cars with an electric ax while wandering the back roads of Beltsville, Maryland. A variation of the legend portrays the Goatman, a retired hermit who lives in woods and is often seen alone along Fletchertown Road nightly.

The Goatman traditions originated "long long, dear, long ago," with the first Goatman sightings being reported in 520 CE as satyrs within Greek mythology. A 1971 incident in which locals claimed the Goatman caused the death and injury of a dog led to them gaining even more popularity. Pearson claims that "bored youths", who repeat the Goatman story and suggest that the creature preys couples who frequent the local lover's street, perpetuate this legend.

Devil of Jersey

There are several variations to this flying biped. Bipedal kangaroos are most commonly described with a horse-like, goat-like, leathery batlike, horned head, horns, short-clawed, cloven hooved legs, and forked tail. It is said that it moves fast and produces a high-pitched, bloodcurdling cry.

Mother Leeds, a Pine Barrens native, created the Jersey Devil. Folklore claims that Mother Leeds, who had twelve children, cursed the child for being pregnant the thirteenth time. Mother Leeds was found in labor during a rainy night 1735. She was surrounded at all times by her friends. The infant was born as a regular baby. But it soon transformed into a creature sporting hooves that were bat wings, bat wings, a goat's head, and a forked neck. It growled and screamed, and it flew its tail up to everyone, before taking off into the forest. Mother Leeds was a witch within certain categories of the tale. The father of the baby, however, was the devil himself.

Sightings

Numerous sightings were reported of the Jersey Devil.

Stephen Decatur was looking through the Hanover Mill Works' cannonballs, when he noticed a flying creature. Joseph Bonaparte claimed he saw the

Jersey Devil hunting on his Bordentown Farm in 1820.

The Jersey Devil was blamed in many cattle kills in 1840. Similar attacks in 1841 with tracks and cries were also reported. Unidentified animals attempted to steal Greenwich farmer's hens. He said that it was impossible for anyone to recognize it from the hundred that had seen it.

Reporter for Pennsylvania Bulletin on July 27, 1937 linked a mysterious critter, with red eyes, to the Jersey Devil. Gibbstown youths from New Jersey claimed they had seen a "monster," which was similar to the Jersey Devil.

The Jersey Devil was reported in hundreds of newspaper reports from the state, ranging from the week of January 16 to 23 1909. The beast was reported to have attacked both a Camden-based trolley car and a social clubs. The creature was shot at by police from Camden and Bristol (Pennsylvania), but not injured. Other stories started with unidentifiable footprints found in the snow. Sightings of Jersey Devil-like creatures were found in South Jersey as well as Delaware and Maryland. In the Delaware Valley, panic set in after the widespread news coverage. Many schools were closed and many employees stayed home. According to some rumours, the Philadelphia Zoo offered a 10,000-dollar bounty for the monster. This incentive has sparked hoaxes such as a kangaroo that had fake claws or bat wings.

Mothman

The Mothman (a winged, bipedal humanoid) is a bipedal. He is not a Mothman despite his name. Instead, he resembles a large humanoid Owl. His color ranges from brown to black, but he is more often seen in darker hues. Mothman is about seven feet tall with a wingspan ranging from ten to fifteen inches and the ability to fly at speeds exceeding 100 mph. Mothman can sometimes be seen without a head. His chest is adorned with two large red eyes. These eyes are said glow or reflect light.

Original Story

Roger Scarberry and Linda Scarberry, two young couple from Point Pleasant West Virginia were driving late at night through an area called the "TNT region," home to a former WWII bombing plant. Linda noticed two large, glowing eyes in the dark near the North Power Plant. They soon realized that the eyes belonged 7ft tall, with its wings folded over its back. Roger stopped and examined the creature briefly. The four men realized that this wasn't a common bird and extended their wings to chase the creature along Highway 62 up to Point Pleasant City.

Sightings

A number of witnesses witnessed similar sightings following the initial sighting. Two volunteer firefighters witnessed the sighting and described it to be a "big bird, with crimson eyes". Mason County Sheriff George Johnson believed that an exceptionally large heron was responsible for the sightings. Newell Partridge, a contractor said that the monster appeared to be looking at him from the spotlight in a neighboring area. His eyes sparkled "like bike reflectors," and he attributed the noises he heard and the disappearance and buzzing of his dog to the creature.

The number of sightings increased over the next year. Mothman was first reported in the Point Pleasant Register on November 16, 1966. The headline said, "Couples View Man-Sized Bird... Creature... Something." Unnamed Ohio newspaper editors later gave him the name "Mothman".

The sightings stopped in 1967 with the collapse on December 15, 1967 of the Silver Bridge. This resulted in 46 deaths. John Keel's book, The Mothman Prophecies claims that the sightings of Mothmans by locals were premonitions of bridge collapse.

Sightings are not restricted to West Virginia. Mothman sightings worldwide have been reported. Some conspiracy theories state that he was present at Chernobyl in the aftermath of the tragedy. This includes when planes struck World Trade Center building and the 1978 fall in Freidburg (Germany). According to Svobodnaya Gruziya, Russian scientists claim that UFOlogists saw the Mothman (a Russian figure) in Moscow prior to the 1999 Russian apartment explosions.

Some believe Mothman seeks to warn people about impending tragedies. While others believe that he is a harbinger and bringer of doom.

Skunk Ape

The skunkape is an humanoid creature similar to the Pacific Northwest Sasquatch. It has a shorter stature and long patches of fur on its shoulders and arms that are similar to an orangutan. The creature is described as having light skin around eyes or faces and resembling a gibbon. It is so named due to its foul odor and appearance.

The skunkape, also known by the swamp cabbageman, swamp apes, stink apes, Florida Bigfoot and Louisiana Bigfoot as well as Myakka and swampsquatch and Myakka-skunkskunks apes, has been a part folklore throughout Florida, Georgia and Alabama since the arrival of the settlers. Esti Capcaki in Seminole mythology is an equivalent foul-smelling, physical powerful, and secretive creature. This literally means "cannibal great". One of Florida's earliest reports of a large simian being dates back to 1818. A report originating from Apalachicola Florida described a mansized monkey or an ape that was destroying food stores, and following fisherman.

Sightings

Skunk-ape sightings were common in the 1960s as well as 1970s. In suburban Miami-Dade County of Florida, sightings were made in 1974 of a big, hairy, foul-smelling, apelike creature that was standing tall on two feet. A number of eyewitnesses saw the creature in Florida's suburbs. In 1977, a Florida state representative introduced but failed a law prohibiting the taking, possess, injury, or molestation anthropoids.

Alien Animals: Top Five Troubling Cryptids

1 - Alien Big Cats

If you've never met a Big Cat or Alien Big Cat, chances are that you know someone who has. The United Kingdom and USA should not allow cats larger than dogs to roam their countryside. However, there are many eyewitness stories that prove otherwise. The reports generally follow a pattern: They occur in a rural area and the spectator freezes, while a big cat like creature sprints away. The presence of what appears be a panther would cause one to freeze.

The most plausible explanation is that many sightings of exotic cats occur because they were released into nature by owners who misunderstoodly believed that they would make good pets.

2 - Big Foot

Big Foot, as it was shown in the Patterson–Gimlin movie, has been associated with America. However there is a similar phenomenon worldwide. Yeti (Sasquatch), the Abominable Winterman, and Sasquatch were all spotted in different cultures and countries. Big Foot is a large, man-like giant ape which can be seen frequently from afar in remote regions of North America. The Sasquatch, while not a wild gorilla, is an important part of the human race. They have fundamental skills that are similar to early man's, suggesting the existence at the very least of one evolutionary branch.

There are many theories and explanations that explain the hominids. This type of cryptid has inconclusive footage. However, most of the information we have is rumor-based and folklore.

3 - El Chupacabra

This Latin-American cryptid, known as "The Goat Sucker", is dangerous to animals and could pose a threat. Though descriptions may differ, it is believed to measure approximately the same size as a child and look similar to a lizard. It also moves like a Kangaroo and has a body that resembles a kangaroo. It has been described as having many Devilish characteristics, including red eyes with an offensive odor and red eyes. A symbiotic animal is one that has a recurring odor problem.

Although it might exist, the creature isn't widely believed to pose an immediate threat to humans. Mexicans are advised to be calm and keep their goats secure, but they can also protect your children.

4 – Monkey Man of Delhi

The Monkey Man of Delhi may not be a familiar tale, but it is a collection of linked eyewitness accounts. New Delhi saw a bright-red-eyed, four-foot-tall hominid with unfavorable disposition and bright red eyes. A cryptid was alleged to have attacked and escaped individuals, leading to injuries, deaths, and even scrapes. An artist's portrayion of this creature did not make it available for examination.

5 – Mothman

The Mothman is another odd creature with glowing red eyes. This cryptid is 7 feet tall and has wide, moth-like wings. Although the Mothman is undoubtedly a larger phenomenon than others, he is still related to Point Pleasant, West Virginia. In the town, there were numerous sightings between 1966-1967 of a tall and winged figure. It presumably terrorized and disturbed many previously normal and well-informed people. Sightings of the Mothman of Point Pleasant increased in intensity after and during the fall of Point Pleasant's Silver Bridge. This led to widespread suspicion that Point Pleasant's Mothman was a

mysterious phenomenon that appeared
before tragedies.

Chapter 2: Bigfoot – Myth And Reality

Bigfoot is often called "Sasquatch" and is the most sought after creature in cryptozoological circles. Bigfoot is an ancient mythology with a history that spans centuries and is still controversial. The Sasquatch (a supposed apelike creature) lives in forests. It is often described to be a bipedal, man-like, big, hairy ape. The mysterious creature has been reported to be seen and photographed from all over the world. Sasquatch (which translates to "wild man") is a British Columbian Indian dialect. "Bigfoot," the most common moniker for the monster, was created by journalists in the nineteenth-century to describe large footprints associated with the creature.

Bigfoot has been described differently in different places, with some slight variations in the details. Bigfoot is an ape-like creature, usually between 6-15 feet tall, 400-600 pounds, and covered in brownish to reddishbrown hair. The creature is commonly described as having large eyes and a low set forehead. It also has a prominent line at the brow. People who claim to be able to smell the creature's odor often say they were able to see it. The tracks have measured anywhere from 24 to 8 inches in both length and breadth, giving rise to the term "bigfoot."

Sasquatch sightings are not new. Reports of them have been made for millennia, and they continue to be reported until today. The Samish, Lummi Klallam, Salish tribes and Lummi have long held legends about the creatures of the Pacific Northwest. These stories tell of a creature that looks remarkably like the Bigfoot seen in modern times. Modern sightings have provided a wealth of information about the creature, including photos, film, plaster casts and plaster casts. It is also possible to use hair samples to test for DNA. Patterson-Gimlin's movie is undoubtedly one of the most well-known bigfoot movies. Its legitimacy continues to puzzle scientists to this very day. The film shows the creature stumbling along "Frame-35323," which depicts a Bigfoot creature. This footage has been seen by many scientists, picture experts, and fraudsters.

Also, there have been several plaster casts containing alleged Bigfoot prints that were made in different parts of the world. Many Cryptozoologists believe that these casts are crucial evidence for Bigfoot's existence. Many casts that are well made show evidence of a pushoff mound in middle of footprint. Many scientists think this trait is valuable for its ability to prevent duplication. This evidence of footprints also provides interesting information about human comfortable steps distances. They can vary from approximately half to slightly more than an individual's average standing height. Sasquatch's steps are over three feet. This means that the steps would be very difficult to imitate or artificially create by someone wearing fake shoes. Many researchers claim that many eyewitness sightings were made of the creature, most often in areas where there are rivers, lakes, or streams, and also areas with high rainfall. This suggests that the creature is not merely sighted out of curiosity, but rather is an ecological niche.

According to scientists, Bigfoot could have been referred to as "Gigantopithecus" by some scientists. Fossils belonging to this monster were discovered in China. It is also believed that many animals have crossed over the "Bering land bridge". It is therefore reasonable to suppose that Gigantopithecus prospered, which would explain some reports of bigfoots. This theory is largely a speculation. There is not enough information available about Gigantopithecus so it's difficult to make a valid link. Sasquatch is a mysterious primate that lives alone in remote wilderness areas. Some believe that Sasquatch is an UFO-produced primate.

Scientists will continue their debate about the existence Sasquatch. Many people believe that it is a mix of myths, folklore and simple misidentifications. Most people believe these sightings to be hoaxes. The Sasquatch continues to be popular and many groups have been set up to try and track down its existence. These groups spend endless hours documenting sightings of the Sasquatch and looking at the data from various locations around the world. The Crypto Zoological profession has made the Sasquatch their favorite creature over the years.

The Legend of Bigfoot -- The Sasquatch

Many have claimed to have seen a huge-haired wild creature in the woods while hiking the Northern Pacific region in America or Canada. The creature is said to stand between 7-8 feet tall, although it may reach as high as 10-12 feet in certain cases. People who have seen these animals say that their bodies are covered in long, shaggy fur. Consider the scent that Uncle Rufus leaves behind after a bender multiplied with ten.

This creature is sometimes called Bigfoot. However, its northern neighbors prefer Sasquatch. Bigfoot, what is it? Take into account the size of the sleds needed for someone eight feet high. Naming the feet is far safer than naming any other bodily part.

Bigfoot is often dismissed without hesitation by those who are well-respected in the scientific world. They will dismiss the hundreds-year-old stories, footprints and films of Bigfoot as frauds committed by bored mountain men. It is hard to imagine how they would entertain themselves in such remote areas without any TV. They inspect hair samples and determine if they belong to a recognized species or an unknown species. Oh, those nighttime noises. You can be sure someone has accidentally stepped on one the fire logs out in the woods. You would too scream.

Every year, credible individuals are ridiculed for reporting sightings going back to when Europeans first arrived in New World to explore it. Native Americans were afraid to see man-like, enormous creatures wandering naked through woods. How do we hold them accountable for their actions? We would be terrified of a large, hairy and nude animal running through our yard.

The body is where the debate's most serious section is. Why isn't a fossil or body ever found? Bigfoot should not have been on the roads already. It's not hard to imagine that you could collide with an eight foot-tall beast on the side of the road. Advocates note that it is very rare to find the remains large creatures still alive in the woods. This is because they are slow to decay in the forest. Bigfoot could bury their sibling and sister in the same manner that people do.

Bigfoot proponents claim that fossil records exist of the large-sized creature. Most will point at the history of Gigantopithecus to claim that this creature came from Asia. Science doesn't want scientists to have to go out into the wilderness in search of these creatures.

Bigfoot and Cryptozoology

Bigfoot has been sighted across the United States since the 16th Century. Native Americans were the first ones to witness Bigfoot. Various Indian tribes referred his existence by different names. 'Sasquatch" was one of 50 Indian names. This creature appears to have the appearance of a hairy male, rather than an ape monkey. Although the creature is often described as tall at 6-8 feet, many people have seen it with a smaller woman and its offspring. The adult males of the species have a rough, muscular build. It has broad shoulders, a large neck and relatively small neck. Its body is covered by short, black hair. However, the hair on its head is much longer. Most sightings of the animal show that its eyes glow or shine in the dark when lit by a flashlight.

People who encounter bigfoot generally report no odor. But those who do experience it may have noticed a strange anomaly. Bigfoot can emit a scent via other means than spraying. Some people have reported that they can smell one thing and then pick up something completely different. There are many odors that can be described as rotten flesh. Bigfoot may utter ape-like grunts or growls, almost to the point where it can be heard screaming. Other witnesses have reported hearing whistles and strange sounds.

Bigfoot is believed, almost unanimously, to be an un-Physical creature. Many Indian tribes claim to having witnessed the creature change into a wild wolf. Some people believe these creatures exist on a plane other than the physical and can travel around the world whenever they please. Indians believe

Bigfoot has extraordinary psychic ability. The creature can also be seen by certain people while being invisible to others. Many stories have been told by non-Indians who witnessed the creature in response to a UFO encounter. Bigfoot is believed to be a spiritual creature, according to some who have spent decades searching for and researching it.

Great Lakes Indians warn anyone who walks through woods hearing a stick strike a log or tree that it is Sasquatch territory. This intriguing observation has been reported by both Indians and non-Indians. Witnesses have reported that the stick hitting can sound louder than normal, like a large log is being struck against trees, or smaller sticks are being used. Several Bigfoot researchers claim to have seen Bigfoots after they were exposed to the strange sounds. Numerous Bigfoot researchers report rock throwing as an additional fact. This is in addition to stick striking. Researchers and their cars have been struck with stones. Bigfoot hotspot residents have also reported stone throwing on their roofs, homes, or cabins.

This may be the strongest evidence yet of their being spiritual entities. Think about your paranormal research cases that include 'poltergeist" encounters. In this case, you will quickly see that a large number of them involve the hurling of stones on homes and the roofs of houses where the poltergeist attack occurs. Much like the bigfoot cases were, the contact or infestation of many poltergeist victims was started by the stones being thrown and dropped.

Another link to the spiritual realm can be found when sticks are struck against trees and logs. I was very small when I saw an advertisement in a magazine. The advertisement read like this: "Connect with God with Amazing JuJu Sticks. The JuJusticks were carved from a sacred bit of wood and sanctified through voodoo. One was able to hear tappings, or other forms from the spirit world when the sticks were struck against wood objects or tables.

The Michigan Dog Man Creature Cryptid & Sightings

Many Wexford County residents became witnesses to the Michigan Dogman, 1887. According to Michigan legend, it is. According to legends, the creature canine-like but with a masculine head and seven feet in height.

This bipedal creature can be seen with amber or blue eyes. It howls in a similar way to a human scream. It is most common in the northwestern part of the lower Peninsula, but reports have come in from other parts. Legend says that the myth describes it as appearing on a ten-year cycle. Years ending in figure 7 are its years.

Steve Cook helped to increase the popularity of The Michigan Dogman in 1987. It was due to his involvement in creating a song that described the creature and detailed sightings within Michigan.

History

According to legends, it roamed the Manistee River where the Odawa tribes called home. However, it had remained mysterious for a significant period of time. It was discovered by large swathes of modern-day humanity in the late 20th century.

In 1887 two lumberjacks in Wextollford were the first ones to meet the creature. According to their accounts, the creature was a man's corps with a dog's head.

The following incident occurred in the Paris area, Michigan in 1937. Robert Fortney was the victim. Fortney said that one of the wild dogs that attacked Fortney included a dog that had two legs. Other sightings were reported by Allegan County residents in the 1950s. In 1967, similar entities were seen in Cross Village as well as Manistee.

The Cook Song

The Cook Song was inspired by Steve Cook's 1987 single "The Legend". The song was written in this time period while Cook was working at WTCM FM Traverse City, Michigan.

Cook was unaware of Michigan Dogman at the recording time and the song's first performance was an April Fool's Day hoax. Cook claimed in a statement he wrote the song out of his head before learning the story.

Cook remains skeptical of the possibility for a genuine dogman. Cook replies by emphasizing that imagination can be the fountain of folklore, just as his song does. He continued by saying that he trusts witnesses who claim to have seen this creature.

Cook credits Bob Farley as providing the music's keyboard backdrop. Listeners phoned in to recount their encounters with the exact same creature shortly after the song played. The station also became the station's most requested song in the weeks following the debut airing.

Cook began selling cassettes in the four-dollar range, with all proceeds going to animal rescue. MonsterQuest featured the creature on its March 2010 episodes. The Monsters and Mysteries in America Season 2 episode, Great Lakes: Wolfman, Dogman, Wendigo, featured the creature in January 2017.

Cook added new lines to the song after hearing about an intruder canine. A mandolin was added to the 2007 recording after the 1997 modifications.

The Michigan Dogman: Darkly Fascinating Stories

Realism is what makes Michigan Dogman an urban legend. Bigfoot is the most well-known cryptozoological creature, with many sightings. Although it is believed that this creature has a supernatural origin there are many animal-based explanations. The first sightings were in Michigan. They continued to spread across the state.

While descriptions may vary, the fundamental characteristics of the Dogman are consistent. This is normal for any species developing, migrating, growing or changing.

Dogman sightings can be seen in the woods near logging camps or on remote roads late into the night. The legend of the mythological beast in backwoods folklore became very popular after Steve Cook's 1987 hit song, "The Legend" (translated from Traverse City radio). Cook drew lyrics from Dogman encounters. There were many more sightings shortly after. But what are these people seeing? Is this mass hysteria? Or the presence of an actual creature? Consider the history of this mysterious creature and the numerous terrifying encounters that it has provoked.

The Michigan dogman is exceptional in his leaping ability

Ray Greenway had just returned from Manistee Army in September 1986. It was nightfall and he saw something in the dimly lit area. Although it appeared that his headlights were reflecting off the eyes, they were way too high above him to be eyes.

The unidentified, unidentified monster suddenly approached him. He leapt straight across the two lane road. "It is impossible for any animal to be it. Ray later remembered that the creature was not a deer. He went on to describe the creature's bright blue eyes and its amazing leaping ability. "I did recall seeing both eyes, as though they were looking at my face. This experience, together with my leaping abilities, will always stay with me."

1887 marked the First Time that the Dogman was sighted.

In 1887, the first recorded encounter between a Dogman and a man was made. Two lumberjacks came across a monster while out walking in the woods. They described it as having a man's body, but a dog's head.

Additional sightings started to trickle throughout Michigan's Upper Peninsula. Residents discovered dog tracks within the mud surrounding several of the horses killed. According to reports horses died because they were afraid.

A Dogman screams like an infant in the night

One witness reports that late fall 2001 saw a strange creature, best described as either a Dogman or a werewolf, stalking the hill behind their home. "My stepdaughter (and I) were gazing out the French door when we saw something that looked black and resembling an enormous bear with wolf like haunches.

Cass County's residents, Michigan, saw this image, but it was not the last. They can still hear it swimming in the swamp on nights when it isn't there. Sometimes, they hear it shrieking. They claim that it screams like an infant. "It's hysterical. It is loud.

OnStar captured a Dogman Incident, May 2006; the notorious Dogman OnStar incident occurred in Troy in Michigan. A man was driving down the road and was suddenly confronted, he said, by a "huge large" dog standing up. He instinctively steered away from the beast, and the result was that he ran off of the road and tipped the vehicle onto its sides.

He was uninjured, and he and his passenger made it home safely. OnStar helped him summon help. This OnStar recording quickly went viral. Watch it in the video.

The Michigan Dogman scouts the streets in ten years.

While the Michigan Dogman's description is changing over time, we can assume that it's not a lonewolf. They are often brightly colored with icy blue or amber eyes, and their heads have terrible fangs. They can stand on two feet and are six to seven feet tall. Legend has it that these creatures are fast and they have a horrifying scream-like howl.

According to local tradition, the Dogman rises in ten years cycles. However, sightings by the Cryptic were initially reported in Michigan but quickly expanded to Wisconsin. Soon after, reports about alleged sightings were circulated across the country.

Robert Fortney Launches A Shot Towards A Smiling Canine

Robert Fortney, 17, reported to police that he was approached by a large and black dog near Paris, Michigan's Muskegon River. Fortney did report the incident 49 years later. The beast rose to confront him and looked down at him with piercing eyes.

Terrified, he shot at the thing which flew away swiftly. Fortney was nevertheless shaken and remarked on the incident over the years. "It could have been fear. But I swear, the dog was smiling at my face."

A 13-year old girl was trying to smuggle a cigarette, and the Dogman caught her.

Courtney, 13-year-old girl from Reed City (Michigan) in 1993, was the subject of a memorable Dogman story. She decided to go outside her family's residence one winter for a quick cigarette, and she received the scare that would change her life. She said she noticed a light glinting between planks of an old abandoned building and that it caught her eye.

She followed its trail and realized gradually that something was there. She saw that the creature measuring six feet was in the barn with a dog's head, and it was staring at her. She fled, fearful. Afterward, her neighbor confirmed that she too had seen a monster outside the barn. They described it as a big dog, about the size and shape of a buffalo, who was prowling the barn.

The Dogman's speed is amazing and he seems to take pleasure in darting in front cars

Rhonda, from Three Lakes in Michigan, had a Dogman encounter back in 2009. Rhonda and her son were traveling on US-41M with their son when the beast bolted onto the highway near Tioga Creek.

"We were perplexed. It might have been a mutant or wolf. However, we couldn't understand the sightings." She said, "The beast ran very quickly and was approximately 50 yards ahead of our truck." It was astonishing because its front was substantially larger than it back, which is larger than a dog.

She claimed she had never heard of the dogman before this encounter. After studying online descriptions and accounts, she can confirm that what she saw was a Dogman.

Dogman enjoys looking at people at night and staring down

Ron, a guy from Lansing Michigan, stopped to help a stranger who was driving along a remote road. He believed he was seeing a deer. "Suddenly, an almost human-like hand appeared over the hillside. It was much larger than a normal man's. This revealed the enormous silhouette. From the dark hillside emerged an enormous figure with a likeness to a wolf.

"It looked much larger than I expected, and had a wolf-like head and eyes that were reflected in the headlights. It sat there, and then turned its face to look directly at me.

Dogman is extremely clever

Chuck initially thought he saw a mountain lion approaching Nestoria Road. Chuck was sure he saw movement near the road's edge in 2011, despite the fact that it was cloudy, dewy, and the weather was a bit chilly. He stopped at a crossroads, waited, and nothing happened. When he was able to run fast enough, the thing reappeared.

"It leapt across the road from the south side to escape the trees within seconds. The fact is that it ran three times before it vanished.

He witnessed the creature in shock. He had never seen anything similar. It was on all fours. It had the back legs a man could have. It was jet black with a human-like head and legs. It didn't have a tail. It has the profile and head of a dog. It didn't make any noise."

Perhaps The Beast Of Bray Road Is The Dogman

Michigan Dogman, a bipedal humanoid with a dog-headed head, was the first to be observed. It is possible that the same species may have different names in different parts. Wisconsin's Beast of Bray Road, a case that most eyewitnesses consider a werewolf, is perhaps the most famous. Some people have compared it with a Sasquatch. Others describe it as a bearhybrid creature or giant primordial Wolf.

According to the Beast of Bray Road, it has both human characteristics and wolf attributes. As the Dogman, an adjacent cryptid to it, the Beast of Bray Road enjoys long jumps. It enjoys leaping across Bray Road in the night to scare vehicles.

There are many theories that support its origin.

It is possible that these creatures exist. The logical answer to this question is that they may be a wild species of canid. Many believe the Dogman may be a primordial, wolf-like creature similar to an Amarok.

These sightings may be legitimate. However, it is possible for species with unusual sizes to be misunderstood or multiple creatures incorrectly to be classified as the exact same thing. This is why most people describe the creature as wolflike. However, occasionally it appears to be a large and malformed member in the bear family.

Chapter 3: Mothman - Alien From Another World, Or A Being From The Spiritual Realm

Mothman, a film and the 12-foot-tall stainless steel sculpture on display in Point Pleasant West Virginia, is more than just a film. It is a strange creature, first seen in West Virginia near Ohio's border, between November 1966 & November 1967. Robert Roach painted the creature as people saw it in sculpture. It is a man-sized creature that has wings and large reflective eyes. The creature's dimensions and peculiarities suggest that Mothman is a paranormal phenomenon. This was most likely to have occurred between 1966-1969.

Mothman's visit to the moon is linked to other alien visits and a new form of Alien. Mothman appeared on November 12, 1966. His name was inspired by the Batman television series. Five men were making a grave at Clendenin Cemetery, West Virginia, when the first apparition happened. According to workers, the "brown human figure" with wings flew over their heads while they were constructing a grave in a cemetery near Clendenin, West Virginia. Two young married couples from Point Pleasant saw the sighting three days later while out on a drive. They noticed two red lights at the shadows the local World War II TNT Factory.

After they stopped, they saw that the lights were the burning eyes of a massive animal. It is described as being about the size of a man with six-and-a half to seven feet in height and large wings folded against its back. After spotting the creature once more on a ridge above the road, the thing chased the couple and forced them to flee in their car. The attack on one couple was confirmed by Mothman after the meeting.

Mothman appearances happened in West Virginia throughout November 1966. Following these claims, individuals in black appeared around the area. One year later, it was unknown who the Mothman and the men in black were. It is not known where they came from. Some West Virginians believe Mothman came from an alien race and that the mysterious visitors were government officials investigating strange occurrences. But the majority of West Virginians believe Mothman actually was a paranormal creature. The mysterious visitors were Catholic priests en route to exorcise the demonic creature.

What is the Mothman and why are we so obsessed?

What characteristics are essential for a good American modern myth? You can't get it out of Google. Something that pervades all those affected and the communities that have grown up around them. It was a lingering entity that manifested itself again and again, confusing the memories of the first time that it came out of the darkness of nothingness to make its way into the world.

Perhaps the most widely-respected and talked about myth in modern times is that of The Mothman. He was an evil monster discovered late 1960s. Since then, the Mothman is often seen in the United States. According to some, the Mothman is a sign for imminent calamitous happenings. Others claim it is an extraterrestrial being with ties UFOs and Men in Black. Many believe it to be a hoax.

The Mothman has a long history of appearing in American mythology. These pages provide a detailed explanation of the Mothman's origins, their connection to a horrible small-town tragedy, and a plausible explanation why we can't seem to stop talking about them.

Where did the Mothman legend start?

Mothman sightings began at Point Pleasant, West Virginia, in 1966. Five workers digging in Clendenin on November 12 reported that they saw a dark-colored apparition resembling the figure of a man fly above their heads.

Two young couples, Roger Scarberry & Linda Scarberry, and Steve & Mary Mallette told authorities on November 15 that a black creature 10-foot in wingspan with glowing red eye sockets was following them in their vehicle. This occurred also in Point Pleasant and close to "TNT Area," an anti-war weapons facility.

Over the next 12 months, more sightings were reported. On November 16, 1966, his first newspaper appearance was in Point Pleasant Register. It featured the heading "Couples Observe Man-Sized Bird... Creature... Something." Later, an Ohio newspaper copy editor gave him "Moth Man," sounding suspiciously like Batman.

Many believed the Mothman hid in a nuclear power plant that had been shut down near the town. This is a location where a government facility was used to test nuclear weapons. The Mothman was this a result state tampering? Is it a winged manifestation due to weapons testing or state tampering? It was the result of wild imaginations that created mythology.

Jersey Devil - Fact Or Fiction?

The Devil and the Pines

This artwork of the Jersey Devil depicts a red-winged Jersey Devil in a foggy forest scene.

To understand the legend of the Jersey Devil's birthplace, it is important to first comprehend its history. It is a vast, unpopulated area that covers about 1700 square miles in the southeast of New Jersey. It is surrounded in dense stands of white cedar by a huge aquifer. The interior is still and quiet. The swamp's waterways get their Tannen dyed by the cedar tree lines. The Pygmy Forest is an area of stunted tree. Many people view it as a desert. However, it is home of 27 orchid species. Cedar swamps made travel difficult early on. Some highways were built by Indians. Others are remnants the stagecoach era. Some roads are paved and others are made out of sand. Roads lead to Hog Wallow. Double Trouble. Sooy Place. Mary Ann Furnace is another location. These names date back to colonial days when the first settlers arrived here in New Jersey. The Jersey Devil calls the Pine Barrens home.

The Origins of the Devil

Leeds Point is the subject of one famous story. A booster woman gave way to a child when it rained on a night in 1735. The room was dimmed by the flickering candlelight. The wind howled. Some people believe that she was an occultist. Mother Leeds, the destitute mom, was said not to have had more than ten children. According to some, the boy was born malformed. Some believe that she cursed her child because of this. Some reports state that the child was born normal and developed strange traits later in life. These included an extended body, wings shoulders, cloven hooves a thick tail and cloven hooves. Folklore claims that the child was kept prisoner until it could escape either through the cellar doors or up the chimney. The Jersey Devil is a creation.

Jersey Devil image in black and White

Another story describes the love affair between a young Leeds Point girl, and a British soldier. The British had originally arrived in the region to supply privateers iron from the Batsto furnaces. In 1778, at the Battle of Chestnutneck, the British defeated America. The match was condemned by the residents of the area, who called it treasonous. They cursed her. Legend says that the Leeds Devil was named after her child.

An alternative version of the story tells about a young woman who stumbles upon a passing Gypsy and begs for food. She was afraid and declined. She was terrified and refused to accept the curse of the gypsy. The girl was blessed with her first child in 1850 when she gave birth to a boy.

Another interesting version: An American Mr. John Vliet was entertaining his children in October 1830 using a mask made by him. A huge face mask. It quickly became a local tradition and was soon accepted by the people. It quickly became popular and was repeated late October, when children and parents alike wore terrifying costumes and faces.

The Chupacabra & Cryptozoology

The Chupacabra, or Goat Sucker as it is often called, has been spotted in South America by thousands of witnesses. It has been detected in South America in Puerto Rico, Mexico, and the United States of America in the middle of 1990 and present. This creature is a mix between an alien called the 'greys' or a porcupine. It has large red eyes that shine the same as fire or lights, gray hair, fangs, spine quills and feathers that may act as wings. It can stand on three-toed shoes and is about 4 to 5 feet in height.

Close encounters revealed that the Chupacabra emits a foul-smelling, pungent odor similar in smell to sulfur and rotten yolks. It has a grunting sound that resembles a wild boar, pig, or wild pig. It can make loud hissing and even sob simultaneously. Since its initial sighting, hundreds of animals have been killed by this creature, including cats, dogs and chicks.

The creature entered a Caguas home through a windowsill and tore into the stuffed toys of a child. One time, the creature made a change in color and a horrified witness saw it. The creature then began emitting strange hissing sounds. The witness became confused and weak after the creature started hissing and quickly vanished.

The creature was encountered by a policeman who heard a hissing sound. The creature entered a house through an Arizona front door. He then murmured and gestured towards the homeowner before retreating. A few hours later the man heard his son screaming and saw the creature sitting on his son's chest. Through an open window, the creature flew from the bedroom. Some witnesses have said that the monster can fly even without wings. There are no obvious ways of guiding it or propelling it. Sometimes, it can disappear suddenly.

The sticky substance the creature left on his floor sounded suspiciously like ectoplasm. This gooey slime is commonly seen during ghost sightings. This is the classic succubus/incubus attack posture described in an account that the creature was found on a boy's chest. The succubus can be either a beautiful, or a terrifying female demon known as the night hag'. While the incubus a masculine form, the succubus can also manifest as a feminine or male demon. According to reports, the creature can appear or disappear instantly, suggesting the existence and existence of a supernatural being or spiritual being. Also, the creature's ability for flight, not soar and glide, without flapping it wings, defies aerodynamic laws. This suggests that something supernatural is possible.

The Sal'awa, a Creature of Cryptozoology

This cryptid can be seen in Egypt's Sohag and Luxor regions. Locals refer to this creature as a similarity to a dog, or wolf. Sal'awa is a strange creature believed to frequent large sugarcane farms in the vicinity.

Legend has it the Sal'awa is an animal that looks like a canine and may reach five feet in height. It is long with a long tail and has big pointed ears. Shahba Hamzeh Shaker from the region has witnessed the sal'awa. He described it as being larger than a doe, but smaller then a camel, with a long neck. This creature, despite its small size, is thought to have a sharp sense of hearing and an ability to blend in with the environment. This mysterious sal'awa hunts lizards as well as other animals. Some residents believe it hunts people as well.

Egyptians call the canine beast Al-salaawa. The animal has been blamed for a lot of the horrific deaths reported in the region. The most famous Sal'awa attack was in 1996, when a pack made of these dogs attacked both animals and humans. One sal'awa was also shot to death by some locals.

Interviews with police revealed that the most recent attack on a 7-yearold boy by the vicious beast occurred in October 2009. The mother of the child believes that she was in a position to protect her child because she distracted the monster. The mother bought a dog to defend her family members from the terrifying beast.

Many Egyptians, tourists and locals fear this beast. These people can be blamed by some skepticals. After all, they were the ones who witnessed this creature in person.

Sal'awa can also be related to Egyptian mythology by being a Set animal. Seth is the Egyptian god, who is feared for his ability to cause chaos and darkness. This set creature has been popular since ancient times, when it had been feared for causing violence and pain. This is also evident in hieroglyphs. This creature is described with squared ears, canine bodies, canine bodies, and a forked trail. The similarities between the Set animals and the sal'awa ones suggest that they are related.

A Sal'awa Information Centre has been established by Egyptians in Armant. It provides additional information about the critter. It contains images and additional information about the sal'awa.

The Human and Animal

Recent DNA analyses have shown that humans share the majority of their genes with other animals. Our physical similarities to other animals far exceed our differences. However, the Western perspective makes a clear distinction among humans and other animals. Since they could not communicate in their language, it is thought that humans have little in common with other animals. Only humans have a soul according to Westerners. We also have a wide range of emotions. Despite our cognitive dissonances, we continue establishing intimate relationships and anthropomorphizing the animals near us. Many societies hold a completely different view about humans' place within the natural world to ours.

Although beliefs differ greatly, many people believe humans are closer to other spiritual and bodily creatures. I'll start here.

Take a look at several of these non Western ideologies and their conceptions of the human-animal connection. Compare them to Western ideas.

Many societies that had historically animistic religious beliefs shared the concept of an era when people were animals, and vice-versa.

Animals could take on the human form of "Distant Times," "Dreamtime," "lithotomy" or other terms. It is said that animals used to display human characteristics.

Many cultures believe that souls may still exist even though people can't see them.

Charles L. Edwards is a folklorist who believes that this notion may have evolved from a memory about an earlier stage in humanity's history when both humans & apes had one common progenitor.

The apelike creature lived the same life as all other predatory animals that inhabited his habitat. However, some descendants of him began the process for transformation later in his life.

...adaption would result in our species' emergence."

Instead of fighting with his adversaries, the diverging elementary person began to strategize.

"As an instance

Edwards believes early humans began to see beyond what was around them and begin to look at the unseen.

"[T]hese powerfuls took shape in the gods that dwell beyond the clouds and cosmogony as well as transformation tales developed." Now,

Adherents of animistic rituals are looking for explanations for the observed events and ways they can tie their rituals into larger processes.

They recall the days when myths were made, when humans were closer than today to other animals.

Edwards establishes the link between the profound spiritual communion with other creatures and the formative stage.

Childhood is a stage in the development of a person. He talks about his childhood and the times he spent watching ants in his garden and inventing.

Stories that reflect the exploits or "the anti-people." He thinks of them as soldiers that work in different industries during peacetime while showing "extraordinary military prowess" during wartime.

Children's imagination is their unique asset. It is the ground on which miracles can grow.

Through the anthromorphism of joyful storymaking, you can feel a sense de metamorphosis as well as a greater sense of connectedness.

The infant is a witness to this as well, just as we did in prehistoric time.

People's [sic] projection of fears, love, or desire into the environment that becomes embodied."

Many non-Westerners consider the traditional rituals that are linked to storytelling and traditional practise an extension of childhood. It is a place where wonder and innocence rule.

Children's contact with nature in childhood leads to a deeper understanding of the world and other forms. However, Western adults are mostly not like this.

We are, on the surface, eager to move beyond childhood. Our deep desire to be a part a larger community of living beings is evident in many different ways. For example, how we feel about our companion animals.

The Koyukon peoples Distant Time myths, which can be found in central

Alaska's boreal forest, are another example of how humans and different species are interconnected.

Animals living in cultures that are not Western. The human-animal period of metamorphosis into animals is once again seen as a dreamlike moment in the earth's genesis.

Distant Time stories feature natural phenomena as well as occurrences. A myth influences how the myths are told.

You must be careful. Koyukon believe that, because animals have been previously human, they are able to understand and be aware of human thoughts, words, acts and actions. Despite

Although spirits of certain animals may be stronger than others it is vital to treat them all with respect, as they can bring you sadness or bad luck.

Otherwise. It might look like this because the skin is identical to other Distant Time animals.

They don't believe there is a distinction between the human and animal realms. The Koyukon believe in a distinct soul for humans and animals.

The spirits and souls of animals. But, they believe humans were created in the image of a Raven. This makes it less significant than Western societies.

The similarities that humans share with other animals do not stem from their animal natures as much as animals' human natures, which have been humans in Distant Time.

Inuit and Eskimo mythology are centered on the idea of relative delimitation between the animal and human realms. They are both stories that span a comparable time.

They represent their worldviews and provide guidance for their customary observances.

Koyukon. They also explain the creation of all life forms.

Determine what each person must do to fulfill their responsibilities in society.

Tom Lowenstein from Tikigaq examines the Inuit's response to this issue.

In a poetic novel called Ancient Land, Sacred Whale. These people consider the annual whale hunting and the meticulous preparations it requires to be a highlight of their year.

Imagine a mythological story. The complex interactions between the rituals associated to the whale hunt and the whale's spirit, which is seen as more powerful, are evident in the rituals.

Superior to those of human. Their belief system allows for the synthesis and blending of many opposites including human-animal. Raven Man also had the doppelganger.

Because the whale embodied both bird and human characters, the uliuaqtaq [an unemarried woman who marries Raven Man] was a dual creative/destructive presence.

The two main elements of the animal-land distinction were identified. "The Tikigaq Inuits play an essential role in preserving this age-old tale by reenacting the story each spring.

You can help ensure that the natural elements are under control and you have a better chance of securing your life and livelihood.

The existence or the Sea Mother is a central part of many Eskimo groups located in northeastern Canada as well as Greenland.

A creature that lives below the ocean floor. It also has a soul which lives in marine animals. All these cultures share the ancient myth that she ascensioned to the title of the spiritual monarch undersea. It unites the animal realms and the human. According to one,

According to myth, Sea Mother (known by many names but Sedna is the most well-known) was once a young lady living with her dad. She had declined.

To marry. But a seabird disguised to be a man convinces her otherwise, and she is whisked across the sea. Her relationship with him ends in misery, and her father abandons them.

He comes and joins her in his vessel. The birdman becomes angry and causes a storm that causes the boat of to sink. The woman is left to cling to life by her teeth. In a fit, rage and despair, her father abandons her, amputating her fingers. Each finger transforms into a marine creature when it falls into salt water. After the last finger is severed,

The woman goes to sea and becomes the Sea Mother.

It is vital that the Sea Mother is happy for the Eskimo, who rely heavily on sea critters as their main source of sustenance. She is considered to be possessive.

Dominion can have control over the souls many animals. They are capable of taking on both animal and human forms as well. Her authority is seen as more than that of humans, because humans are entirely dependent on other animals for their survival. She is however disregarded for her refusal of joining human civilization. This is evident in her refusal not to marry and her determination to live in fantasy worlds. The Sea Mother's role as both an abject and a mother is impacted by the human/animal split.

They regard her as an outcast. She is powerful and should be feared. If they want to survive, however, they must find a healthy equilibrium.

Chapter 4: The African Waterbobbejan

South Africa's Waterbobbejan has been accused in terrorizing people and even causing the deaths of a few. Waterbobbejan terrorizes livestock by ripping off chickens, goats, cattle and other animals that it can catch. Waterbobbejan's name means "waterbaboon" and is described as being pygmy-sized at approximately 2.13 meters or 7.6 feet. Its fur could be described as being as dark as scorched Earth or red. Numerous eyewitness sightings have been made of this creature, ranging from the deep woods up to the outskirts African capitals.

Jaffe has been interviewed as the world's most respected expert on the subject. Jaffe, the world's foremost expert on the subject, stated that he had studied a genetic line taken from a tuft of Waterbobbejan hair. The sequence proved to be identical to that of a human, except with an extra chromosome. The fur itself was nearly as thick and dense as a dog's. Jaffe also claims that the fossil record does not contain the Waterbobbejan. There are many gaps that could be filled with fossils that could have contained the Waterbobbejan. Jaffe stated that the Agogue in the village was also recognized as a Waterbobbejan picture. The Waterbobbejan was shown as an oversized man with a pointed nose, unusually long arm lengths, and thick black fur. Jaffe, however, stated that this alone is not convincing and it doesn't account for a solid fact. There are no historical documents on the South African Waterbobbejan. This could indicate that the history in South Africa might be a clue to the missing documents.

South Africa has a long past, especially with the official apartheid system that lasted the majority of the 20th Century. Apartheid saw the minority white population control the government and enforce segregation of the different races in housing, education, and just about every other aspect of life. Three nations were created as a result. The first of these is the whites. These are people primarily of British or Dutch ancestry and have struggled for generations in order to gain political supremacy. It culminated with the South African War 1899-1902. The second is that of blacks. These include the San hunter/gatherers of northern desert, Zulu sheders of eastern plateaus, as well as the Khoekhoe farms in the southern Cape. The third country is comprised of mixed-race persons and ethnic Asians. These include Indians Malays Filipinos, Chinese and Filipinos.

60 people were killed and 180 injured in 1960 when police attacked a non-violent demonstration organised by the Pan Africanist Congress. Sharpeville massacre was the catalyst for activists moving to more militant tactics. Nelson Mandela was made the leader for the military wing, which is part of the African National Congress. In 1961 they launched

a campaign of sabotage against government installations. Mandela and the other members from the African National Congress was later detained and sentenced in 1962 to life imprisonment. Nelson Mandela, who was being tried in 1962, spoke out for freedom, democracy and equality for all South Africans. Nelson Mandela was sentenced to a long time in prison and has come under increasing international condemnation.

The 1960's saw the emergence of the "Homeland System," which was designed to bring about the end of apartheid and create independent homelands for blacks. These were poor, rural areas that didn't have the ability to function as separate states. It was then that 13% of the country split into ten homelands. 80% of the population lived in the homelands.

Most of the world community disregarded the apartheid systems and

South Africa was ranked among the "pariah" states. By the mid-1980s, South Africa had been the target of economic and cultural boycotts. Mongane Waly Serote, a South African poet during this time, said the following:

"The country's poor have an urgent need for self expression. When I speak of self-expression, it doesn't necessarily mean people speaking about their own selves. I'm referring to people who are making history. We often neglect the creativity which has enabled people to survive extreme exploitation, oppression, and death. People have survived extreme race. It means that our people have been creative in their lives.

It was necessary to create a new administration in order to allow for changes. President FWDe Klerk was reelected as president in 1989. He lifted the bans that had been placed on the ANC or other opposition groups. De Klerk and Nelson Mandela were able to secure the South African transition to democracy under difficult circumstances. They had to contend with different opinions from both tribal and politically-oriented factions in the black community as well as opposition from certain whites. De Klerk and Mandela came up with a common solution and Nelson Mandela was elected the President for South Africa in 1994. Nelson became the leader of the Government of National Unity. In this government, the minorities like Klerk's National Party's vice president were represented.

The new government quickly recognized the need to act if unity was to occur between the divided communities. The Truth and Reconciliation Commission, (TRC) named Archbishop Desmond Tutu to its leadership in an historic attempt by the TRC to define the apartheid era's violence and human rights abuses. The commission heard the testimonies from both victims and criminals over three years. The Truth and Reconciliation Commission, despite not being intended as an instrument for punishment, was able witness to the sufferings of victims and could grant amnesty to any perpetrators. The 1988 findings of the commission were made public by the commission. All social groups were condemned for the atrocities they committed, including the activities of the Mandela United Football Club (led by Nelson Mandela's ex wife Winnie).

Truth and Reconciliation Commission's vision healed the wounds left by a horrific past by acknowledging and accepting the truth. South African's willingness to engage in the process won them respect across the entire globe. Nelson Mandela called upon the nation to be proud of their actions and to take steps to improve them as a nation in response to the findings. There are many instances where South Africans were tempted to work in mining, as part of their long history.

Although there are evidence to suggest that small scale gold mining occurred in

South Africa's greenstone belt areas prior to the rise of modern gold mines, very little information is available about this past. South Africa's recent history of gold-mining began with the mining of greenstones in northern Kwazulu-Natal around 1836. This was followed by the development and operation of mines in Murchison, Giyani, or Pietersberg. Nearly forty decades later, in 1875 the first discovery of gold was made on the farm Kromdraai which is located north of Krugersdorp. This lead to the Witwatersrand area's first official proclamation of gold. Barberton saw the discovery of the Pioneer Reef in 1883. Meanwhile, the Witwatersrand Main Reef proved to be very valuable in 1886. South Africa welcomed miners from all walks of the globe. New companies were established and the country's gold mining industry was developed. South African has 342 tonnes of gold, making it the largest of all international gold mining corporations.

AngloGold Ashanti Gold Fields, Harmony Gold Fields, Harmony DRDGOLD and Western Areas all have large gold mining operations in South Africa. These five companies accounted, together, for 91% of all the South Africa gold mined.

AngloGold Ashanti currently runs seven mines in South Africa. The Vaal Region is located in the North West Province near Klerksdorp and includes three mines: Great Noligwa Kopanag and Tau Lekoa. The West Wits Region is second, near Carletonville. It contains three mines - Mponeng Savuka, TauTona, and Savuka. Gold Fields is also based in Johannesburg, South Africa. They also have operations in Australia as well as Ghana. Harmony's corporate address is in Johannesburg. Harmony's headquarters are in Virginia in Virginia. DRDGOLD's headquarters are also in Johannesburg. It is listed both on the JSE Limited, Port Moresby, Australian, London and Australian stock exchanges. DRDGOLD had 6,390 employees in South

Africa by the end June 2005. Western Areas and Placer Dome have a combined 50% interest in the South Deep coal mine. In 2004, the South Deep mining operation produced 13.4 tonnage of gold. 6.7 ton of that was attributable by the Western Areas. Western Areas employs a total 4,914 people, as of the 31st Dec 2004

The rich history which should have seen more interactions with the Waterbobbejan - from apartheid history to gold mining history - might offer answers to many of your questions.

Section 1: Zoology

There must be an animal that can be used to illustrate the Waterbobbejan's Zoology. The Waterbobbejan can be most closely related to a species of ape. There is no ape-related monkey found in South Africa.

Waterbobbejan has been the subject of rumors since the 1880s. However, a notable sighting took place in 1965. Two boys, Koster and Swartruggens from the

North-West Province, South Africa saw the animal at the Leeufontein farms. There are currently two possibilities for the Waterbobbejan's identity. The Waterbobbejan could possibly be the Chacma babooon. Chacma babies are well-known but only reach 114cm, or 3'9", in length. The Waterbobbejan could also be called the samango monkey. Although samango monkeys have a smaller size than Chacma monkeys, there has been at least one instance where a farmer killed and injured a samango and claimed that it was a Waterbobbejan.

Waterbobbejan's diet is the same for both chacma baboons (and samango monkeys), except that they are smaller. Chacma baboons eat all kinds of food, including fruits. Chacma babyoons can also be found eating shellfish and marine invertebrates, particularly at the Cape of Good Hope. Samango monkeys have a diet that includes fruits, insects, flowers and leaves. These diets vary from the Waterbobbejan's because it is mostly carnivorous. However, the Waterbobbejan eats some livestock, including cattle.

There aren't any South African animals that could be considered closely related to Waterbobbejan. Waterbobbejan literally means "Water baboon" so chacma and other baboons in South Africa are thought to be the closest relatives of the cryptid. Baboons can be the largest monkey species and are very intelligent.

Baboons use repetitions of certain sounds to convey emotions. This type is the primary method of communication used by the baboons.

Scientists estimate that there are thirty baboons who make vocal calls. These include barks, barks and grunts. Nonverbal parts are also part the baboons communication repertoire. These include their body postures as well as facial expressions. Communication among baboons requires not only a vocal component. Baboons can also display aggression using the so-called "open-mouth threat". This type of threat involves the raising of the baboons' eyebrows, revealing their whites, and then exposing their teeth. Baboons who are more hostile can stand their hair up to give the impression of a larger size.

Numerous studies on several species show that baboons use flexible vocalizations. For example, female chacma babies use a "loud bark" type of call that can be adapted to the caller's predator type and the social setting at the time. Field researchers were able, after studying the acoustic attributes of the calls closely, to discover that there was a continuum of variability in baboon's calls. For example, when a baboon wishes to stay in touch with its group, or when it is worried about its infant's safety, it may give more tonal sounds. For large predators, more noisy calls were made. There were clearly marked differences in alarm call categories between calls made in response and seeing alligators, and calls made in response and mammalian canivores. Both types were quite distinct from the type of contact call received. There were also quantitative, consistent variations in all types and calls for different individuals.

However, Waterbobbejans may not be as social and friendly as baboons. Chacma babies can communicate very well verbally and in nonverbal communication. Waterbobbejans would display more primitive communication features. Waterbobbejan communication seems to be similar to that seen in orangutans.

Waterbobbejan may have a communication method similar to those of orangutans. But, it is different in its feeding. Waterbobbejan's diet overlaps with many mammalian cativores in South Africa. Due to the overlap of Waterbobbejan's diet and the connections between baboons (and

Waterbobbejan), there are many theories.

Section 2 Theories
The first theory on the Waterbobbejan's existence is the chacma babooon theory. Chacma baboons can often be found in South Africa. They reach lengths up to 3'9" (excluding their tails) and may weigh upto 45 kilograms (or 99 lbs).
Waterbobbejan was translated from the original name Waterbobbejan. It is, as we have said many times before: "Water baboon". Another piece that supports the chacmababoon theory is the fact that they are the longest monkeys and consume a part of smaller vertebrate creatures. Papio ursinus (the scientific name for chacma baboons) is derived form the French words meaning "Bearlike baboons". This definition can be used to support Waterbobbejan's description. However, the Waterbobbejan cannot be proven to be a chacma babooon.

Although chacma babies are large, they are almost twice the mass of the supposed cryptid who would weigh 122.5 kilograms or around 270lbs. Chacmababoons do not have the Waterbobbejan's hair which is dark brown to black. Waterbobbejans are not chacma-baboons. This is also evident in their diet (see Section 1). Waterbobbejans have a carnivorous diet whereas chacmababoons tend to be omnivores. Chacma babyoons are highly social animals and can live in groups up to four hundred people. Waterbobbejan, however, are extremely shy animals and can be found alone despite their name meaning one of the most social primates. The second theory concerning the Waterbobbejan's existence is that of the samango monkey theory. A farmer who shot and killed a Samango monkey to prove that it was a Waterbobbejan is the best evidence. There are differences, however. Some sources also claim that there were multiple similar incidents involving other farmers. However, the Waterbobbejan and samango monkeys have many other differences than their overall appearances and habitat. One of the many differences between the two species is the fact that the Waterbobbejan's weight can exceed 50 pounds, five times more than the largest male Samango monkeys at 9 kilograms. Food is another important factor. Samango monkeys, which are carnivores, do not have the ability to eat as many insects as the Waterbobbejan. The third difference is in social activity. Similar to chacma monkeys, samango monkeys are also social creatures, which is contrary to the Waterbobbejan.

Waterbobbejan's behaviour is very similar to that of leopards. Leopards are the most social cat family, being graceful and powerful. The leopard is so strong and at home in the trees, that it often carries its prey along with him into the trees. Leopards are able to drag the large bodies of large animals higher in order to defend their prey against hyenas, who can then eat them. Leopards can also hunt in trees. Their spotted coats make it possible for them to blend into the leaves and spring with a devastating pounce. Deer and antelope are also easily spotted by leopards. Leopards may attack people and dogs when they see humans. Leopards are good swimmers, and they are used to swimming in the water.

Female leopards may give birth at any hour of the day. The cubs are typically grayish and have barely visible spots. The mother will hide her cubs and move them around until they're old enough to play and learn to hunt. Cubs typically live with their mothers until they are two years old, with leopards living alone for the rest of their lives.

The black fur of leopards is similar to Jaffe's illustration of Waterbobbejan's encounter with the village. Because their spots are difficult to distinguish from the fur, black leopards may appear to have a solid colour. Black panthers often refer to leopards with black fur.

However, other characteristics of the Waterbobbejan are not explained by the behavior or color of leopards fur. Waterbobbejans are closer to monkeys and baboons when we consider the factor of their name and the possible explanations for their identity. Waterbobbejans may also be larger than leopards when they include their tail. Leopards can reach 3.28 meters (10'9") in length. The Waterbobbejan on the other hand can reach up to 7 feet in height at 2.13 meters. Jaffe's study with the black fur from the supposed Cryptolith did not match the color of Jaffe's fur.

Waterbobbejans are not supported by biology because they have the fur and bodies of primates and behave like leopards. Furthermore, paleontology disproves Waterbobbejan in the forms of leopards or primates.

Modern leopards, which evolved around 500,000 year ago, are the most adaptable large cats and have the longest history. Diet is an important factor in the rapid, successful evolution of modern Leopards.

A total of 29 studies, including four unpublished, on leopard diet were published. They included estimates of relative prey abundance for the leopards. This analysis was done in 13 countries and forty-one locations around the global distribution of the leopards. A Jacobs' value was calculated to each prey species. Each study was then compared with a median of 0 using sign tests or t for preference or avoidance. According to research, leopards are more likely to eat prey that is 10 to 40 kg in weight. Regression plots showed that 25 kilograms was the most desired weight of leopards' prey. The mean body weight of significantly preferred prey was 23 kilograms. Leopards prefer prey with weights within this range. This is possible because leopards hunt in small herds and in dense habitats which offer little danger of injury. The preferred prey species are impala, bushbuck, or common duiker. Chital could be preferred with a larger number of animals from Asian sites. The preferred species range includes species that are not in the desired weight range, those restricted to open habitat, and those with adequate anti-predator strategy. The ratio of the average leopard body mass and that of their favorite prey is less than one might reflect the leopards' solitary hunting strategies.

Leopards are a catholic in the use of their habitat. These range from tropical rainforests and desert savannas to alpine hills and the edges of urban environments. The riparian zones are home to the largest leopard population. This is because leopards can survive in any place with enough cover and the right prey animals.

Leopards come in a variety of morphologies, with adult leopards weighing between 20 and 90 kg. To maintain their body mass they consume between 1.6 - 4.9 kg of meat daily. In order to meet this daily food requirement, leopards hunt around 40 prey animals annually in the Londolozi Game Reserve near Kruger National Park. This includes fifty in Kruger National Park while they kill sixty in the Serengeti. The leopard's average body weight is 21.5 kilograms. Leopards' large body mass could allow them to survive for short time on small vertebrates or other invertebrates in areas where there are few large prey. Accordingly, leopards are known to prey on small species such a birds, rodents (hares), catfish and up to the size giraffe calves or adult male eland. The sub-Saharan Africa record for leopards includes 92 species of prey, which is the largest number of large predators. However, the records are mainly focused on the 20 - 80 kilogram range.

Leopards are mostly solitary. However, the females' territories are generally enlarged by the larger territories of the males. In open habitats leopards hunt alone at nights. The camouflage of leopards allows them to stalk close to their quarry, before sprinting up to 120m. The average leopard's sprint time in Kaudom is 10.3 + 1.3 metres. This allows them to run up to 60 kilometers an hour. Contrary to popular belief, leopards in rainforests hunt daily with crepuscular highs. They stalk their prey by ambushing them at fruiting trees or along game trails. The leopards' attempts to kill them end with only 5% success in the Serengeti and 16% in Kruger. 38% of Kaudom hunts ended in defeat. Additionally, between 5 to 10% of leopards kills are lost due to other predators, such as lions. These losses are compensated by similar levels scavenging. The leopards prevent kleptoparasitism (a term that describes parasitism through theft), by hoarding carcasses. Caching is often used to protect carcasses, but 57% in Kruger were tree cached carcasses. This was especially true for spotted and spotted hyenas. Kaudom had only 9% carcasses that were dragged into thick Kaudom foliage. The strength of leopards can be seen in the records of giraffe calves found cached in trees.

Jacob's Index, one of many methods to minimize biases towards large prey species, can be found in several studies. Jacob's indicator uses r to denote the kills by a single species at a location, while p indicates the prey abundance. The result can range from -1 up to +1. Where -1 represents complete avoidance and the carnivorous creature completely avoids prey, +1 indicates maximum preference. Prey animals are considered to be one of the most predisposed prey. +-n where the number is between -1, +1. The mean is +-n. Five animals might be close to the Waterbobbejan's preferred prey: the puku; the water chevrotain; the red forest double-eyed duikers; Thomson's gazeelle and the African cuvet. Jacob's Index shows that leopards favor the puku, while water chevrotain are 0.82+-0.17, red forests duikers prefer 0.37+-2.21, Thomson's gaze are 0.14+-0.21, while African civet preference is -0.06+0.42. It is possible to calculate the proportional prey most often by using the same method.

Waterbobbejan is a different species than leopards, as they have 20% overlap in the most common prey items. Waterbobbejan isn't just a leopard. The cryptid cannot climb trees like leopards.

Both leopards as well as monkeys have the ability of climbing trees. However, the latter seem to be more popular for this skill. Samango monkeys are perfect examples of arboreal monkeys because they are South Africa's only native arboreal monkeys.

Samango monkeys only inhabit a few forest habitats. The ongoing loss of forest habitat and fragmentation means that samango monkeys only live in isolated, semi-isolated forests fragments. There is a possibility of low dispersal rates due to this. The estimated area of samangos in the wild is more than 20,000 square kilometers. The area for occupancy is defined as the natural habitat that remains within forest patches greater then 1.5 square kilometers.

Samango monkeys live in the canopy of evergreen forest, and their current distribution shows a very wide tolerance for forest habitat. As South Africa's only forest-dwelling Guenons, the samango monkeys have been associated with high-canopy Evergreen Forests. They live in various indigenous forest types, including afromontane forests as well as scarp and coastal forests. Samango monkeys have been seen in human-modified habitats including residential gardens, pine plantations, campsites, and residential gardens. But, it is not yet clear if samangos are capable of dispersing between forest patches using modified landscapes. They are capable of using infrastructure provided by humans to move across their habitat. For example, they can travel along telephone and electric lines as well as roads. Samango monkeys, however, seem to consider human inhabited areas to be more dangerous than their natural habitat. Instead, they prefer to forage within indigenous forests if offered experimental patches in both forests or gardens.

Samango monkeys eat a lot of fruit, and are considered primarily frugivores. Between 50 and 70% of their diet is made up of fruit. However, their main source is from leaves and insects. Samango monkeys eat other plant parts, like flowers and buds, when there is a shortage of fruits. Samango monkeys can eat exotic species of plants, including those that have been planted by humans. Therefore, it is possible for samango monkeys to be considered pests.

The Waterbobbejan, unlike the samangos, does not need the same adaptations. Waterbobbejans live in open spaces, so they are not as restricted to forests like samango monkeys. The main factor that led to the evolution in arboreal life of the Samango monkeys was their diet. This is dramatically different from the Waterbobbejan.

There are many problems because of the differences in diet, behavior, appearance, and other characteristics between the cryptids (and both leopards) and primates. Waterbobbejan was believed to be the culprit for the death. This requires an analysis of the current status South Africa's livestock industry.

South Africa's livestock are raised according to their climatic conditions. The western 67% region of South Africa is very dry, and is only suitable for livestock. Modern, highly sophisticated, and intensified systems to raise livestock are paralleled with subsistence, communal pastoral methods.

Through long years of working in the livestock sector, many indigenous species and breeds have been conserved. Their adaption and preservation is a unique advantage, and it is important to preserve the breeds in order promote extensive livestock production. Breeds do not only serve as breeding materials, but also provide an important contribution to the national genome pool. Many breeders and crossbreeds have contributed significantly to the development in South Africa over many years.

There are more sheep than cattle or goats. KwaZulu-Natal is the exception. The sheep still outnumber the goats (see in). It can be concluded that sheep account for 58% between the three livestock. Cattle account for 28% and goats 14%. This means the sheep outnumber the goats and cattle by 2:4 and 1:4, respectively.

This is an important fact because the Waterbobbejans eat a lot less livestock. We have analyzed the rate of change in South Africa's livestock commodity production between 1972 and 2002. The decreases in commercial productions are mutton (61%), wool (56%) and butter (80%). Condensed milk decreased by 52%. Skimmed milk powder decreased by 30%.

Since Waterbobbejan interviews in the 1860's indicated that farmers had lost most of their sheep, and cattle and other livestock were also lost, the data that was analyzed is vital to the theory. Farmers have stated that the livestock died in a way that is not consistent with natural predators like leopards and lions. A prize bull disappeared with the rest, according to one instance. The farmer went looking for the prize-winning bull the next day and found its remains dismembered. There are stories that a massive ape like creature was seen leaping from enclosure walls with terrified animals underneath its arms. Another story is that many farmers had their fruit trees destroyed by an ape-like creature. Tswana farm labourers believe that the Waterbobbejan dwells in caves, behind waterfalls or near water bodies. They claim that the cryptid will take its prey into a cave to devour it and then tear it apart.

If Waterbobbejans claimed that sheep are their prey, then the number of sheep would decrease compared with the goats and cattle. However, this is not the truth as the ratio between sheep and animals is 14 to 10. These facts can lead us to conclude that the farmers' claims about the sheep and animals are at most partially false. Waterbobbejans hunt differently. The process of dismembering prized bulls is very similar to humans. But, it is also similar to lions eating their prey. The lack of evidence is making it difficult to establish the origins of the hunting method and whether it was different from any other known animal.

Another reason the hunting method could be similar to a known animal is that farmers can be unreliable. Robert Brain, a famous paleontologist and head of the Transvaal museums in Pretoria, examined Waterbobbejan reports. He found that only one incident confirmed the fact that farmers don't provide reliable information.

Dr. Robert Brain received the phone call of a farmer in Mpumalanga who claimed to be the one who had shot, killed and skinned Waterbobbejan. Dr. Brain visited the farmer to learn that the trophy was actually of a Samango monkey. Dr. S. found out that the farmer had exaggerated the entire encounter. Brain, Dr. Brain identification. He believed that he had indeed murdered a Waterbobbejan. This might not be applicable to all farmers, however it should still be noted that the farmers who made the statements about the killings lived back in the 1860s with far less knowledge of biology and forensic science that what we know today.

The Waterbobbejan is not confirmed by reliable evidence, so the cryptid could just be a hoax.

Chapter 5: The Australian Yowie

The Australian Yowie may not be as well-known and loved as Bigfoot's other Bigfoot variants, save for the African Waterbobbejan. Two types are the Yowie cryptids in Australia, according to historical accounts.

The first species will grow between 1.8-3 meters or 6-10 feet in height and can weigh up 454 kilograms or 1,000 pounds. This species is said look like a man-sized ape with talons, instead of fingers. Yowie is considered to be a more human-like species than the North American Bigfoot. The Yowie's face and head are thought to be closer to those of a primat, as well as the ability to walk upright. It is described as being more aggressive, dangerous and hostile towards humans than its North American cousin. It is shorter, measuring between 1.2 and 1.5m, or 4-5 feet in height. Some believe the Yowie might be an ancient hominid that has not been extinct. Local cave art depicts hominids standing tall and hairy beside smaller Aboriginal characters. For the possibility to be true, we must first understand Australia's history and geography.

Australia is an Australian island continent, located between the Indian Pacific oceans. Australia is the sixth most populous country on earth, covering 7.686.850 square kilometers. It boasts a varied landscape that includes rainforests, deserts and snowcapped mountains. Uluru also known as Ayers Rock or the Great Barrier Reef is two of Australia's most iconic natural features. Most Australians descend from immigrants from Britain and Ireland. The native Aborigines account for only 2%, according to the 2013 census. This is approximately 452,000 indigenous Aborigines of the 22.6m citizens.

The Cambrian period 540 million-years ago marked the beginning of Australia's geological history. Major changes occurred during the early Cambrian Epoc. It is important to make a comparison between this era and the Vendian epoch 620million years ago in order to fully appreciate the difference.

The Earth of the Vendian Epoch 620 Million Years Ago was very unusual in comparison to what we see today. A large part of what is now the Pacific Ocean was covered in land. However, the Pacific Ocean covered the region where Europe (Asia) and Africa are found. Two major continents ruled the Earth during the Vendian epoch. However, they are not so well-known as Northern or Southern Gondwana. Northern Gondwana contains what are now called India and Antarctica. Southern Gondwana comprises what are now called Africa (the Americas) and parts of Asia. Presently, tropical areas, such as West Africa or parts of South America are congregated near the South Pole during Vendian Epoch. They were extremely glaciated.

It was approximately 540,000,000 years ago that the Cambrian Era began to show the stark differences between Vendian and Cambrian epochs. The two major continents that made up the Vendian Epoch had merged to temporarily form a supercontinent known as Pannotia. In the early Cambrian, Gondwana, a portion of Pannotia that stretches almost from poles to pole was discovered. Gondwana was comprised of what are now China, India Australia Antarctica, Africa, South America, and Australia. Laurentia and Siberia were two landmasses important that weren't part of Gondwana. The long journey northwards was made possible by a growing mid-oceanic sea ridge that connected the two islands.

The northern hemisphere, half a century ago in the Late Cambrian Epococh, was nearly empty. With the exception of the submerged remains of modern Russia near North Pole, Near the South Pole lies Avalania. It is made up parts of Britain, Ireland, Spain and the eastern American coast, Iberia. Armorica, which consists of other Western European remains, was also found underwater off Gondwana's coast. Eastern Australia was located off the coast of Gondwana. Eastern Australia also contained a series mountains belts that were formed when the continent's old core met with thin silvers at the continental shelf, which are called'microcontinents.' These microcontinents began around 500,000,000 years ago.

At the end of the Ordovician Epoch, most continents experienced a rapid increase in their land area. Volcanic activity intensified the process, adding land to Australia's east coast along with parts of Antarctica. Parts o Gondwana fell apart during this period. The most durable parts moved to the south where North Africa overtook the South Pole.

The majority of the continents formed in the southern half of the earth during the Silurian Epoch 420 millions years ago. Gondwana (which included South America. Africa. Australia. India) was located near South Pole. Avalania (a continental fragment which covered most of the eastern coastline of America) closed an ocean known to be the Iapetus Ocean near Laurentia. Laurentia comprises the majority of North America. The Rheic Ocea began to open south of Avalania. Massive volcanic eruptions took place in eastern Australia between the mid–Silurian and the Devonian periods.

360 million Years ago, in the Devonian Epoch, two huge supercontinents were slowly moving towards each other. Gondwana, made up of South America, India and South America, was situated in the south. Laurentia, a region that consists primarily of North America or northern Europe, is to be found in the North. The American Midwest was submerged by the shallow seas. Iberia - which includes Spain and Portugal - is located off Laurentia's southern coast. The Devonian epoch saw the continued growth of mountains in Australia. The southeastern coastline of Australia

collided in a chain volcanic islands and formed a chain mountains. From the mountains, new rivers carried sediment to basins on the continent's center. This created distinctive Devonian stones. Western Australia was covered by shallow waters. Over time, the sediments from the seafloor have been compressed into the distinctive reef limestone/mudstone. These rocks contain fossils that include incredibly diverse communities and species of fish as well as other marine life.

The Carboniferous period began 354 millions years ago. Pangaea the supercontinent can be seen for first time in geological historical history. Pangaea emerged from the Laurentia. The Laurentia consists of North America as well as Europe. Gondwana's clockwise rotation had occurred before the collision. So the eastern Gondwana which was comprised of India, Australia, Antarctica and South America moved southward and the western Gondwana which was made up of South America (and Africa) moved northward. Gondwana's turn opened up an ocean in the east called the Tethys Ocean and closed down the Rheic Ocean west. At the time, the ocean between Siberia (Baltica) was shrinking, which allowed for another collision. The north of Australia was covered in warm waters in the Carboniferous age. The coral reefs can be seen as limestone at the moment. Dense forests in eastern Africa laid thin layers on coal. Gondwana moved north during the Carboniferous Epoc. There was a cooling of the climate and evidence of this change is in the southern Australian glacial deposit. Australia moved southward by twenty degrees in 30,000,000 years.

Permian lasted from 248 to 290 million years ago. Pangaea formed when Siberia, the ancient island continent, joined with the rest the major landmasses. The crustal plates ran from pole to pole. As Pangaea finished forming the southern continent, Gondwana began to be broken up. An active ocean line opened, creating a new ocean surface. Parts of Gondwana's edges were pushed away from Pangaea, as the Permian-era sea widened. Micro continents that formed and traveled northwards this way included places such Tibet and Malaysia. The Permian age ended with the greatest mass extinction ever recorded, with 90% dying. The dramatic drop in sea levels, and the massive lava flows that erupted in Siberia caused many species to disappear from land and shallow seas. It is estimated that only 5% of marine animals, and 33% of terrestrial creatures have survived. The massive lava flows that formed in Siberia may have contributed to climate change.

In the Triassic era, a large ocean dominated one part of the world while Pangaea was dominant in the other. Pangaea comprised what is now North America. Europe, North Asia. South America. India. Australia. Pangaea moved slowly northward during Triassic. Since the Permian period's end, sea levels had increased again, allowing slow recovery of marine species, such as corals. As climates became more hot and drier in the future, tropical coal-forming rainforests and swamps declined.

Earth was different and warmer during the Jurassic period. It was 170 millionyears ago. It is possible that there was no ice cap at the poles for much part of the Jurassic period. Also, the mild conditions resulted from much higher sea levels. This created a smaller area of land and vast, shallow continental waters, which were alive with life. Pangaea started to disintegrate, and it was then that familiar modern landmasses such North America and Eurasia appeared. The North Atlantic Ocean and South Atlantic Oceans both opened up, while Tethys Ocean was closing down.

The Cretaceous period ran for almost 80,000,000 years, between 142 and 65 million years ago. As the land that made up Pangaea shattered, the Earth began to change. The Atlantic Ocean had formed and extended north and southeastern, separating Africa and Eurasia. As new oceans began to form between the Gondwana elements, Africa and Australia started to drift apart 120 million year ago. Asia was unknown. The land to its south, including Indochina and India were still separated islands.

The Atlantic Ocean split the New World from the Old World around 90million years ago. This was during the Cretaceous period. India was excluded from the creation of much of Asia. India was still linked to Madagascar during this time. While Australia was attached at Antarctica, Australia languished in its deep south. The opening to an ocean between the two continents began with the formation the Southeast Indian Ridge. Australia was gradually moved northwards by new ocean floor that was laid on either side. Australia continued to be separated from the other major landmasses throughout the Cretaceous period. It was a drastic change in the theory of the hominid.

Section 1: Hominids

The main features of human evolutionary, also known collectively as 'hominization', have been the physical and cultural changes. As hominization has progressed, differences between anatomy and life have been less significant than changes made in the way people live, how they use their environment, and how they interact with others. Physical developments include changes in posture and locomotion in hominids. Physical changes in hominization include the expansion and modification of the head and pelvis.

Hominids tend to have smaller teeth and jaws compared with more primitive primates. Also, their teeth are thicker enameled. Furthermore, the faces of hominids appear more flattened than those of other primates. The crest on the skull and the ridges over the eyes of hominids has been lost. However, the brains in hominids tend to be larger than that of other members of their body. The brains from hominids were also developed to be more complicated than those of other primates.

Cultural development includes cooperative work, group formation, tool manufacturing, tool building, harnessing fire and making sculpture and painting. Burial rites are also part of the cultural development of hominids. Each development should not be taken as an isolated event. The hands that were not used for locomotion gave rise to greater ability in the production and use of tools. This allows for better hand-eye coordination and brain development. Also, tool usage places great importance on improved child-rearing skills, social organization, as well as communication. A long-term group of people who provide infant care and support their members throughout life is more likely than others to gain, share, or accumulate experience. Intimately bonded social groups are more likely use tools efficiently and to improve their design.

Many vital cultural developments, like language and social structures, don't fossilize. Accordingly, paleontologists are limited in their ability to predict the future cultural developments from the activity records, which can include evidence of burial rites. Interpretation is very important because of the small amount of data available. There is generally a lot of information available about stone tools, due to their ability for fossilization. But, tools made of animal or plant materials, such leather and plaited-fiber bag, are difficult to interpret.

To learn more about whether the Yowie could be a hominid, it is important to know the origins and evolution of primates, mammals, and mammals. Cynodonts are thought to have evolved mammals about 220,000,000 years ago.

Thus, mammals were already an old lineage at the time Australia was separated from other continents. This marked the completion of the Mesozoic Age of the Dinosaurs. It also marks the beginning of the Cenozoic Age of the Mammals.

Hominids first appeared in the Miocene Epoch 6-7 million years ago. This was when Sahelanthropus Tchadensis from Central Africa, Chad, made his appearance. A complete cranium is known about Sahelanthropus nicknamed Tourmai. Additionally, there are fragmentary lower teeth and jaws. At the moment, it is not clear if Sahelanthropus has a bipedal or quadriplegic nature. This primitive species had many apelike features including a small brain of approximately 350 cubic meters. Sahelanthropus has a mix of apelike and humanlike characteristics. This is in line with the belief that the species was formed around the same period as hominids and chimpanzees.

Orrorin tugenensis (West Kenya) was first discovered. This fossil species includes fragmentary arm bones, thighbones, lower jaws, teeth, and teeth dating back to 6,000,000 years. Orrorin's legs were 150% larger that Lucy, an Australopithecus famous specimen, which suggests it was about the size of a female Chimpanzee. The scientists who discovered this species believe that Orrorin was a human ancestor, able to climb trees and also bipedalism. Aiello-Collard 2001 paper, which is fragmentary in nature, has skepticism about the Orrorin claims. A later paper made by Galik et al. 2004 found more evidence of bipedalism within the Orrorin femur.

Australopithecus.ramidus (from fragmentary fossils, which were dated back to approximately 4.4 million year ago) was named in September 1994. Late 1994 found a more complete skull and partial cranium. The species was transferred to the new genus Ardipithecus, which is based on this fossil. This fossil was extremely fragile. Therefore, excavation, restoration, analysis and publication took over fifteen years. It was published by Ardi in October 2009. Ardipithecus was approximately 120 centimeters tall (3'11") and weighed 50 kilograms (or 110lbs). The skull as well as the brain were very small, and comparable to chimpanzees. While Ardipithecus is able to be bipedal on the floor, it was not as adapted for bipedalism like the australopithecines. Ardipithecus could have been quadruped in trees. It lived in a woodland area with patches and forest, suggesting that its bipedalism did not originate in a desert environment.

Many fragmentary fossils that were found between 1997 and 2001 were discovered. They are estimated to be between 5.2 and 5.8 Million years old. However, they were later assigned in 2004 to an entirely new species, Ardipithecus. One of the fossils found was a toebone belonging to a bipedal species. However, an analysis of that bone showed that it was only a few hundred thousand more recent than the rest. So the Ar. Kadabba's position is not as certain as some of the fossils.

Australopithecus.anamensis has been named in August 1995. It contains 9 fossils. They were most commonly found in 1994 from Kanapoi. In 1988, 12 fossils, mostly teeth, was found from Allia Bay. A..anamensis is a species that lived from 4.2 million to 3.9million years ago. The skull of A..anamensis is primitive, while the body shows advanced features. A. Anamensis shares many similarities with older geologically-aged apes in its teeth and jaws. A partial infibia is a strong indication of bipedalism. And a lower humerus seems very human-like. Although the skull is thought to be the same species as the skeletal bone, it has yet to confirm this.

Australopithecus.afarensis, a type of Australopithecus, was alive between 3.9 and 3.0 million years ago. A. Afarensis possessed an apelike face. It had a low forehead with a bony bone ridge over the eyes, a flat nasal structure, and no chin. Their protruding jaws had enormous back teeth. A. afarensis's cranial capacity varies from 375 to 575 cubic centimeters. A. afarensis cranium is comparable to that of an chimpanzee except for its more humanlike teeth. The canine molars are smaller than that of modern apes. However, they are longer and more pointed that those of humans. A. afarensis has a more human-like pelvis, and leg bones. A. Afarensis can be easily identified as bipedal. However they were more adept at walking than running. Their bones demonstrate their physical strength. Australopithecus. afarensis displays sexual dimorphism. Males are much larger than the females. A. afarensis had a height range of approximately 107cm, or 3'6", to 152cm, or 5'. A. Afarensis' toe and finger bones are longer than those of humans and are curved. A.farensis has hands that are very similar to humans. This is a strong evidence that A. Afarensis has not lost its ability to climb trees. However, some scientists believe it to be evolutionary baggage.

Australopithecus Africanus was alive for about 3 to 2,000,000 years. A. africanus is bipedal like A.farensis. The former has a slightly larger physique. A. africanus had a brain size between 420 and 500 cubic millimeters. Despite having similar body sizes, the brain of A. africanus is slightly larger than that of chimps. However, it is not yet advanced in the areas necessary to speak. A. africanus's front teeth were slightly larger that those of A.farensis. A. africanus is much more similar to human jaws than human teeth, even though their jaws are considerably larger than human teeth. A. africanus is a fully parabolic jaw, similar to humans. A. afarensis' canine teeth are even smaller.

Australopithecus.garhi (from a partial brain) was named April 1999. A. garhi is different than other species in its features, especially the size of its front teeth and primitive skull morphology. A. garhi might be related to some nearby skeletal materials, which have a humanlike ratio for the humerus/femur. Bat an apelike ratio for the lower and upper arms.

Australopithecus sediba was discovered on Malapa, South Africa's site in 2008. Two partial skeletons were discovered. One belonged to a young man and the other belonged to an older female. The partial skeletons can be traced back to 1.78- 1.95 million year ago. A. sediba was discovered to be transitional in A. africanus to the genus Homo. The partial skeletons are more closely related to Homo to any other australopithecine and could therefore be considered as a candidate for the ancestor the genus Homo. A. sediba was a bipedal animal with long arms, which made it easy to climb. It also had humanlike features such as craniums, teeth and pelvis. The volume of the skull of the young male is approximately 420 cubic meters. The length of both fossils is about 130 centimeters. This makes them approximately 4'3".

Each of the above-mentioned Australopithecus species are called gracile Australopithecines. Their skulls and teeth, however, are not as strong or large as the incoming ones. These australopithecines are known for being robust. They were nonetheless more robust and durable than modern humans.

Australopithecus. aethiopicus existed between 2.65 and 2.3 million year ago. A..aethiopicus is based on the Black Skull that Alan Walker discovered, as well as a few minor specimens which may be the same species. This species could have been an ancestor of A. boisei/A. robustus but has a puzzlement mixture of primitive, advanced and basic traits. A..aethiopicus's brain is very small at just 410cm. Also, the skull parts, especially the hind ones, look very primitive, most closely related to A.afarensis. A. boisei shares other features, including a large head, jaws and single teeth.

Australopithecus rigidus had a body much like A. africanus. However it had a larger, more robust cranium. A. robustus lived between two and 1.5 million year ago. The large face has a flat or dished surface with no forehead and large eyebrow ridges. The front teeth were very small but it had large, grinding teeth in the lower jaw. Most specimens have sagittal-crests. Its diet consisted of mostly hard, chewy food. This species' average brain size is around 530 cubic meters. A. robustus skeletons were found with bones that could have been used in digging.

Australopithecus borealis lived between 2.1 to 1.1 million year ago. A. boisei, a closely related species, had more cheek teeth than A. robustus. Some of the molars were up to two centimeters wide. A. boisei's brain size is roughly the same as that of A. robustus, at around 530 cubic inches. Experts concluded that A. boisei was a variant of A. strongus.

It contains many species that belong to the Australopithecus group, making it very impressive. However, Homo, the genus that includes many species from Australopithecus, is even better and has more details than what would be expected to point to Australia's Yowie.

Homo viabilis, which roughly translates to "handyman", was chosen because of evidence found in tools and remains from the species between 2.4-1.5 million years ago. This species was in many ways very similar with australopithecines. H. africanus has a more advanced cranium than H.abilis, but its cranium still projects less. H. Habilis's back teeth, while smaller than the modern human counterparts, are significantly larger. H. habilis brain is about 650 cubic inches larger than australopithecines. The brain size can vary from 500 to 800 cubic millimeters. This is similar to H. erectus at high elevation and australopithecines in the low end. H.habilis's brain shape is also more human-like. Broca's region's bulge (which is necessary for speech) is visible in H.abilis' brain casting. This suggests that H.abilis could be capable of basic speech. H. Habilis was approximately 5' tall at 127 centimeters. He weighed about 45 kilograms or 100 pounds. It is possible females could have been smaller that males.

Homo Georgicus is a type of species that was discovered in Dmanisi. These fossils were thought to be between H. merectus and H. hominis. Three partial skulls and 3 lower jaws were discovered. They all date back approximately 1.8million years. It is thought that the brain sizes for the skulls varied between 600 and 780 cubic inches. H. georgicus was approximately 1.5 meters high, which would have been 4'11" if he had been measured from a footbone.

Homo congenita was a species that existed between 1.8m and 300,000 Years ago. H. erectus also has a face similar to

H. habilis. This includes protruding jaws with large molars. A thick brow ridges is present and a low skull. H.. erectus brain volume ranges between 750-1,225 cubic centimeters. H.erectus has a brain volume of 900 cubic meters in its early specimens. Its later specimens have a capacity of 1,100 cubic cmimeters. H. extinctus has a skeleton that is more sturdy than that of modern people, which means that it was stronger. There are many variations in the body proportions among this species. The Turkana Boy for instance is taller than most modern humans. Peking man has fewer bones and is shorter. H. erectus is more efficient at walking that modern humans. His skeleton reveals this. H..habilis as well as all australopithecines have been restricted to Africa. H..erectus was discovered in Asia, Europe, Africa and Asia. H.. erectus has stone tools that are more advanced and may have used heat, which is consistent with evidence from H.. habilis.

Homo antecessor, a fossil found in Atapuerca Spain some 780,000 years back, was named in 1977. H. thetecessor would then be the oldest European hominid. The antecessor's mid-facial region has advanced features, while the forehead, teeth, and eyebrow ridges are primitive. H. thetecessor is not believed to be valid partly due to its juvenile form. H. sapiens archaic forms first appeared around five hundred thousands years ago. The term can be used to describe a diverse range of skulls, which have both H.erectus-like features and those of modern humans. H..erectus has a bigger brain and modern humans have a smaller one. Also, the skull is more round than H.. erectus. H. sapiens possessed a skeleton with teeth that is less durable than H.erectus's, but are still more sturdy than modern humans. Many still have large forehead ridges and receding chins. There is no clear line that separates the archaic H. sapiens (late H. erectus) and archaic H. Sapiens (archaic H. simpiens).

It is also difficult to classify fossils that are dated back to more than 500,000 years ago as to which species they belong.

The Neanderthal, one of the H. sapiens species, is an example. Neanderthals lived between 235,000 and 30,000. Their brain sizes were slightly larger then those of modern people. This could have been due to their larger bodies. Neanderthals have a longer brain case and a lower skull than modern humans. Neanderthals also had protruding and receding headlights, much like H. erectus. Neanderthals' chin was often weak. Neanderthals also have a prominent middle-facial area, which is something that H. erectus nor H. sapiens do not have. It may be an adaptation of the cold. Other differences between Neanderthals (and modern humans) include minor anatomical variations, the most striking being peculiarities in the shoulder blade and pubic bones. Neanderthals inhabited cold climates.

Because of their short stature and short limbs, their bodies are very similar to those of cold-adapted modern people. Their bones are solid and strong, with evidence of strong muscle attachments. Neanderthals possess many more tools and weapons than H.erectus. Western European Neanderthals often have a stronger form than H. erectus, which is why they are called the classic Neanderthals'.

Homo Floresiensis is a fossil of Homo Floresiensis that was found in Flores (an Indonesian island) in 2003. A complete fossil of an adult female is found at Flores, Indonesia. She measured about 1 meter (or 3'3") in height. Her brain was 417 cubic centimeters. Other fossils reveal that the female was about the same size as the other members of the species. H. floresiensis, a dwarf type of H. erectus is believed to be the species. This is because it is very common for large-sized dwarf forms of large mammals that develop on islands. H. Floresiensis was totally bipedal. He used fire and stone tools and hunted dwarf elephants, both of which are also found at the island.

H. sapiens in modern form first appeared approximately 195,000 years old. The average brain size in modern humans is 1350 cubic millimeters. The forehead rises strongly, the eyebrows ridges, which are usually very short, are often absent, the prominent chin, and the skeleton, which is extremely gracile, are all present. After the Cro Magnon culture was established 40 thousand years back, tools began to evolve. They were made from a greater variety of raw materials including bones and antler. Over the next twenty thousand-years, artifacts such as decorated tools, beads or ivory carvings of people and animals, clay figurines, music instruments, and breathtaking cave paintings became more common.
Long-term trends towards smaller and weaker teeth over the last hundred thousands years can be observed.

Mesolithic Mesolithic Humans, who lived approximately ten-thousand years ago, have a face, jaw, or teeth that are roughly 10% more durable than ours. Upper Paleolithic individuals, roughly thirty thousand year ago, are approximately 20 to 30% more strong than modern humans from Europe and Asia. These are considered to be modern humans, even though they are primitive. Interestingly, aboriginal Australians are the only modern human species with tooth sizes comparable to archaic H. Sapiens. The most small tooth sizes are found in areas where food-processing techniques were used for the longest. This is a likely example of natural evolution that has occurred within the past tentausend years.

There are many options for what the Yowie could become, as there are many species in the genus Homo. The same main flaw which gives European wildmen an edge should be placed within hominid theory: Location.

Chapter 6: The Conclusive Theory

Six types of Bigfoot have been discussed. These include the North American Bigfoot as well as the Himalayan Yeti and South American Mapinguari. The African Waterbobbejan, European Wildmen, and an Australian Yowie. The Yeti. Mapinguari. Wildmen. and the Yowie are all bogus hoaxes of unknown origins. Bigfoot and Waterbobbejan however, are misidentified.

While the historical, scientific, mathematical and historical evidence supports each theory, there is one piece of evidence that needs to be challenged: witnesses. Although there are many explanations of the mass sight witnesses, the main emphasis is on repetition and expectation.

The expectation effects is an effect caused by knowledge of what would happen in the future. This effect is sometimes called the "top-down command". While the repetition effect is caused by sharing property between current and preceding stimuli and is often called the bottom-up priming'. Both effects can be viewed as orthogonal. However, they can be confused. In 2005, Harold Pashler and Liqiang Hua studied the effects of expectation and repetition on the feature values of the target. They also examined the effect of distractors on the defining attributes.

Maljkovic (and Nakayama) previously conducted research on repetition priming, which affects singleton searching. In 1994, subjects searched to find a singleton. For example, white targets have black distractors. Black targets have white ones. Maljkovic was able to prove that even though the repetition of the feature valued was twice as likely than the alternation, the subject's responses were still speeded by it. Maljkovic/Nakayama stated that the priming effect was not caused by expectation but only by repeated behavior. Their study focused only on priming responses to long-lasting displays, rather than the accuracy and perception of short displays. It remains to be determined if the effect was caused by changes or post-perceptual processing.

Huang and Pashler examined this effect by looking at work on responses accuracy using brief displays. In their final experiment they compared accuracy results with the findings for response speed with unlimited viewing. As has been often noted, response-time indexes both post-perceptual and perceptual stages. While accuracy using very brief displays only measures perceptual processing, it does not indicate the latter. Their research aimed to determine the effects on the perceptual stages. Huang and Pashler's studies were primarily used to evaluate subjects' ability to perceive brief displays accurately.

Park, McCool and Prinzmetal provide an example of how response time and accuracy measurements can differ. They are examples from 2003, which looked at the Stroop Effect. The Stroop Effect happens when colors are named. The Stroop Effect occurs when subjects have to respond to the names of different colors (e.g. 'Red'), where the words' meanings interfere with the naming. In neutral conditions, where the words have no relation to the colors the process of naming takes much longer. Prinzmetal et. Prinzmetal and co. did the same experiment using brief displays to verify the accuracy for color naming. They found negligible differences between colored and uncolored words. This result indicates that the Stroop interference is post-perceptual.

Huang and Pashler provided credit for the participation of 50 undergraduates of the University of California, San Diego in their experiments.

The stimuli presented were on a 1,024 X 7,68 MAG MAG X-15T color screen driven by an Intel Pentium IV 1.8G.

Subjects viewed the displays at a distance of approximately 60 cm. They entered responses using the keyboard. Microsoft Visual Basic 6.0 was used and the program was executed on Microsoft Windows 98. Timing routines were also tested using a digital timing device. Each search display had twenty lines. Each line was 1.25m long and 0.21m wide. All lines were white with a luminance greater then thirty cd/m2 and a background that was black with a luminance below 0.2cd/m2. Except for one, all lines had one orientation on the screen. Those are the distractors. The other line, the target, had another orientation. Ten lines were randomly located in two regions measuring 6.65 by 14.36. The regions were placed in the left and right halves respectively of the display. They were spaced approximately 2.13 degrees apart from the center. The correct response to the target location (on the left or right halves of the display) determined its exact location.

Each trial began by presenting a small green fixation in the center. This was displayed for 400 milliseconds. After a 400-millisecond black interval, the display appeared. The subjects were instructed by the instructor to fixate upon the cross and then to search each display for the target. Each display was covered up after a short time. The fifth experiment was the opposite. In this experiment, the displays were left until the subjects replied. The subjects determined whether the target should be in the left or the right half. To do this, they used the j' or k buttons to respond. The left half corresponds to the j' button and the right half corresponds with the button on the right. The tone lasted about 500 milliseconds and indicated whether the response was correct. The next trial began 400 microseconds later. Each subject was able to complete ten blocks out 100 trials. However, the practice blocks were not included. Different blocks were used for each subject, and were counterbalanced between them.

The first experiment consisted of finding and reporting the location a vertical target line within the horizontal distractors. There were two kinds of blocks. In homogenous blocks the feature setting was identical for all trials, no matter if it was horizontal or vertical. A homogeneous block could have all vertical distractors as well as all horizontal targets, or vice versa. The random block had the feature settings randomly chosen for each trial. Homogenous blocks had different feature settings and they were counterbalanced for each subject.

The results revealed that homogenous blocks performed better than random ones, and there was no significant difference between alternate target-featured trials or repetitions. These results suggest that the combination of repetition and expectation about feature settings is sufficient to provide a substantial in perceptual advantage. It is worth noting that repetition does not produce any effect by itself. Either expectation or repetition have significant effects, while repetition has no effect.

The second experiment was set up to allow the subjects to have an expectation of a particular feature value. The target feature values would not be repeated. The task of the subjects was the same as in the first experiment: locate and state the position of a vertical target from horizontal distractors. As in the previous experiments, there were two types alternating blocks. An alternation block was the first. The feature settings in each trial of this block were kept separate from those in the previous experiment. Maljkovic's 1994 study of the role and expectation of anticipation used this type. The feature settings were selected randomly for each trial, in a random block.

Alternate block accuracy was about equal to random block accuracy. Random blocks did not show any clear differences between repeated and alternated trials. Both interpretations could be used to explain the results of the first experiment. However, the second experiment favours the latter. Here, an important perceptual advantage can only be achieved by the operation of both repetition and expectation. For this reason, the priming effect is hereafter called the expectation-repetition effect. A cross-experiments analysis if variance was used, or ANOVA, revealed a significant interaction among expectation and repetition. The same result was confirmed by the second experiment, which had no repetition effect in its random blocks.

In the third experiment the task was to find and report the position of a diagonal distractors target. In a homogenous group, the orientation of all distractors was constant throughout each trial. In a random bloc, the distractors' position was determined randomly each time.

Both types had constant orientations. In the fourth experiment there was a clear advantage in homogenous over random blocks. Target feature values were predictable in homogeneous but randomly in random blocks.

Both experiments proved that homogenous and random blocks were equally accurate. Random blocks had no apparent difference between repeated trials and alternated ones. The third experiment revealed that there was no distractor/feature inhibition effect. However, the results from the fourth experiment show a limited target-feature facilitation. However, this effect approaches significance and should therefore still be considered genuine. Nevertheless, it is considerably smaller than the expectation-repetition effect identified i the first experiment [interaction in across-experiments ANOVA: $F(1,18) = 6.12$, $p < 0.05$]. The expectation-repetition effect can't be explained by either the distractor-feature inhibition or the target-feature facilitation, or by their algebraic summation. Evidently, this effect is dependent upon some interaction between processing the distractor's and target features.

The fifth experiment's purpose was to determine if there was a significant difference between the first experiment, and the research of Maljkovic in 1994. It used short displays with accuracy measurement. The fifth experiment was identical to the first except that displays were present and unmasked up until subjects responded. Additionally, the accuracy was measured and not the response time.

The five-th experiment yielded the results for the mean response time. For homogeneous blocks the response time was 480 ms with an error rate of 2.7%. For repeated trials of random blocks, the response was 548ms with an error of 1.8%. Alternate trials were performed in random blocks with a 548 millisecond response time and a 2.8% error. In homogeneous block responses were significantly quicker than those in random blocs with an effect rate of 91 milliseconds [F(1)= 51.23, P 0.001]. Random blocks saw repeated trials perform significantly better than alternated trials which had an effect at 46 milliseconds [F(1)=51.23, P 0.001].

There is a key difference between the first- and fifth experiments. In the first experiment repetition in random blocks did not have an effect on accuracy. It had a significant effect upon response times in fifth experiment. Given that accuracy was only 0.87% in this experiment, it is unlikely that the 46 milliseconds effect in the fifth experiment could be explained by a difference in perceptual processing. It is possible that accuracy increases slowly over time, as evidenced by the fact that 71% accuracy was reached in the first experimental at a stimulus duration length of 127ms. 71% accuracy would suggest that accuracy was not at its maximum level of sensitivities. The repetition effect on speed reflects a different underlying cause from the perceptual expectation-repetition effect of the first experiment. Both the former and latter are presumably caused by post-perceptual processing, while they both arise from perceptual processes.

Huang and Pashler examined the data from random blocks during the first two experiment. They calculated the repetition effect as a function successive repetitions at the target feature. Huang and Pashler discovered no significant accumulation. They did no analysis of runs that had more than four consecutive repetitions because this type trial is very rare and could not be used to calculate the repetition effect. It is safe to state that the accumulation of successive repetitions cannot explain the expectation-repetition effect shown in the first experiment. It is therefore necessary that expectation be included in their explanation.

This tentative account proposes the notion that a feature divider in featured singleton detection is possible. The feature divider is a mechanism that applies a categorization policy that divides orientation future spaces into two parts. The feature divide's operation causes elements with one set of orientations to be 'highlighted' while those with the rest are ignored. Huang and Pashler tried different methods to find a parameter which could distinguish between one item and another. In the first test, the feature divider could use the exact same parameter for all trials in a homogeneous bloc. However, for random blocks it had to switch parameters back-and-forth. The third and fourth experiments used a single parameter for all trials of a randomblock. This parameter distinguished vertical lines and horizontal lines starting at 45° left-tilted lines.

It is believed the experiment's differing stimulus durations reflect the relative difficulty in the tasks. The difference in duration is not the cause, but rather the result of the expectation-repetition effect. Even in random blocks, this advantage was always available. This is consistent with feature divider account.

If the repetition was an attribute the feature divide that kept the parameter of each trial just prior, then the accidental repetition in the first experiment random blocs should have created a significant benefit over alternation. Because such an advantage has not been observed, it can be assumed that the feature divider does not use a parameter from the last trial in cases where the outcome of changing the parameter could be severe. That is, the underlying mechanism can choose to work in preparation for a given feature or not. This fits with the previous work by Bacon, Egeth and others in 1994 that made a distinction between "detecting an feature" and "detecting one". In the second experiment however, it was found that the random blocks had an advantage over predicted alternation. This could be due to the fact that feature dividers can theoretically hold only one parameter at one time. Replacing the old parameter by a new one will result in the loss.

Naturally, there are alternative accounts of the expectation-repetition effect. Tipper in 1985 stated that the effect could be due to negative priming. This is the result of being unable to attend to an object, or feature, that was previously inhibited. There are two reasons to question the negative-priming argument. First, expectation is an important component of the expectation-repetition effect. It is possible that a feature divider would be significantly affected if voluntary control was exercised, as in conscious expectation. In the case of negative priming, however, voluntary control plays only a small role. A feature's negative priming won't stop just because it lacks expectation. Neill in 1997 said that negative primeing is a reflection post-perceptual aspects. Accordingly, negative priming should not be included in measurements of perceptual precision of short displays.

Moore and Egeth (1998) and other studies can be used to support this conclusion. It was concluded that feature information is not able to instantly improve the perception or perception of any particular feature. There could be many reasons for this phenomenon. Spatial attention can be affected only by feature information. Or, spatial attention mediates featurebased attention. The'mastermap of visual attention' refers to location. By highlighting all locations that include a given feature, search operations can be made easier and are performed in the context of this feature. This idea is frequently expressed in attention models such Cave's 1999 model, Wolfe, Cave, or Franze's 1989 model. Experimental support, such the Johnston-Pashler experiments, also supports this notion.

Huang and Pashler's research seems to contradict Farell-Pelli's 1993 feature-based attention studies, Shih and Sperling 1996, Moore and Egeth 1998. These studies show that feature information does not alter perception in very brief displays. But the present study proves that this is incorrect. An analysis of Huang and Pashler's methods and the differences in their results shows that they have strengthened their concept of'spatial meditation of color-basedattention'. Moore and Egeth should interpret their 1998 results in a way that knowledge about a target features must be translated to location information. This then starts the attention processing for other search dimensions.

There was a significant difference between the preset study and the three other studies. Subjects were required to report secondary characteristics in the later studies, while Huang & Pashler only required them to report location. Moore and Egeth (1998) found that subjects could use the feature division's location map by paying attention but cannot observe its effect because it is slow.

The five experiments lead to three conclusions. Priming can improve perception of feature clues. Priming can improve the perception of feature cues. However, it cannot be triggered by expected repetition and is not triggered just by expected alternation. It is not possible to explain the perceptual increase by either distractor or target feature enhancement. Finally, the expectation-repetition effect reported is different from the repetition priming effect. The repetition priming affect is dependent on repetition only, rather than expecting. It seems to be purely post-perceptual. Meanwhile, the expectation-repetition effect requires both repetition and expectation, and appears to have perceptual components. These conclusions are critical in the case for the cryptid. They show that there is little to no improvement in perceptual perception in nearly all encounters. Psychology supports this piece.

Children learn texture, shape and sight when they reach out to touch things. A pair of kittens from 1963 demonstrated that vision requires body movement.

Richard Held and Alan Hein were two researchers at MIT who placed a couple of kittens in a cylinder ringed with vertical stripes. Both kittens saw visual input by moving inside the cylindrical. The experiences of the two kittens were quite different. The first kitten walked by itself, while the second rode in a gondola attached at a central axis. This setup enabled both kittens to see the stripes move at the same speed. If vision was simply about photons that the eyes receive, then both of the kittens' visual systems should develop in the same way. Surprisingly enough, the results reveal that only the kitten with normal vision developed while walking by itself, while the kitten riding in a gondola never had to learn to see.

Vision isn't just photons. It's more than that which the visual cortex can easily interpret. Vision is more than just a visual experience. Training is the only way to make the signals arrive in the brain meaningful. It involves cross-referencing these signals with the information from the actions and consequences of a person. It is how the brain interprets visual data.

Perception involves the brain comparing sensory data streams against one another. This comparison can be difficult due to the difficulty of timing. Different speeds of brain process all sensory data. A racetrack sprinter is one example. There is almost a tenth of second between the gunfire at the start and when sprinters reach the finish line. Although athletes work to reduce this gap as much as possible, biology is the real limit. The brain must register the sound, send signals back to the motor cortex, then to the muscle cells.

Experiments revealed that sprinters respond to light more slowly than a bang. This is due the speed of information processors. Visual data takes more time to process than auditory. Signals that are flashy take longer to travel through the eye than those with bang information. The light response time took 190 milliseconds. Sprinters, however, reacted in 84.2% less time than the bang signal. They responded in 160 milliseconds.

The alternation is usually activated within 30 milliseconds in most cases. The illusion of Bigfoot, and its cousins, is thus created. Even though this seems unlikely, it is not surprising considering the bizarre reaction time. It is therefore necessary to provide more evidence in order for a plausible hypothesis to be established.

First, certain sensory processes exist. Specialized nerve cells are able to be activated by certain stimuli (light or sounds) and provide the sense experience. Stimulated receptor cell, also called afferent nerve cells, create a chain reaction that excites nearby cells to create an neural pathway. These connector neurons link the brain and central nervous system. For example, signal from the brain travels along the same paths to efferent or motor neurons in order to stimulate muscles, and control bodily movement.

Sensation is the result of the brain receiving electrochemical messages from the sensory systems. Each sense transmits a distinct type of signal. Different brain regions, analogues to different sensory organs' processing of these signals, receive the signals in a variety of ways. For example, information is received from the eyes by the primary visual cortex. The information then gets analyzed by the neighboring vision association and is interpreted. This information is then applied to cognitive psychology to make Bigfoot and its relatives more familiar.

The term "cognitive psychology" refers to the modern approach to psychology that emphasizes mental processes over behavior. Psychology was founded in the early days to study the brain. The cognitive revolution was the catalyst for cognitive psychology. Cognitive revolution is a movement that is heavily influenced by the advances in information and computer science.

Hermann Ebbinghaus is a German psychologist who pioneered scientific studies of memory. This includes cognitive psychology.

Hermann Ebbinghaus grew up in Barmen in Germany on the 24th of Jan 1850. Ebbinghaus began his education as a student at the University of Bonn, where he studied philosophy and history. Ebbinghaus then studied in Berlin as well as Halle. Ebbinghaus' studies in France were interrupted by the Franco-Prussian War. Ebbinghaus enlisted in Prussian forces at 20. Ebbinghaus completed his doctoral dissertation as a philosopher at Bonn's university after the war. Ebbinghaus studied in France and England during 1873-1880. He began to take an increasing interest in psychology. Ebbinghaus taught at Berlin's University of Berlin from 1880-1886. Ebbinghaus had a son in 1885 named Julius. Julius would go on to become a well known philosopher. Ebbinghaus began lecturing at the University of Berlin. He was then appointed professor in Berlin from 1886 until 1894. Ebbinghaus held a chair both in Berlin and Halle between 1886-1894. Ebbinghaus created laboratories in Berlin and Breslau. Ebbinghaus is also the founder of the Zeitschrift fur Psychologie sowie Physiologie der Sinnesorgane. The 'Uber die Hartmannsche Philosophie des Unbewussten' (1873), two volumes in 'Grundziige des Psychologie' (1902) and the 1907 edition of Psychologie' (1907) are some of his most significant publications. Breslau, Germany, was the final resting place for Hermann Ebbinghaus in 1909. He died of pneumonia in 1909 from an infection called "Grundziige der Psychologie" in 1902. This infection can inflame one or both of the lungs and could have been caused by fluid.

Ebbinghaus's most notable observations were his identification of patterns of how people learn or forget. These discoveries set the stage for the current study in memory. Ebbinghaus's methods had less impact than his discoveries. Ebbinghaus believed there was no way to directly access the minds of other people. Therefore, introspection (the examination or observation) of the person's mental processes is the only way to study these processes. Other experimental psychologists considered introspection subjective, unscientific. These experimental psychologists devised other methods to study cognitive process and many other subjects.

Ebbinghaus discovered one thing: two thirds are lost after just 24 hours. Ebbinghaus realized that there were many methods to avoid the exponential and rapid 'forgetting-curve'. Repetition is a way to keep information in your brain. Information also becomes more accessible from the memory due to repetition.

Ebbinghaus's experiments have shown distinct patterns of memory loss and forgetfulness. Ebbinghaus found a similar learning curve for memorization. This is because of the rapid onset and slow decrease that mark the forgetting curve.

Bluma Zeigarnik is a Russian psychologist. While she was visiting her local cafe, Bluma discovered another distinct feature of memorization. If someone asked them about their order, they were able to recall the exact details. However, waiters found it difficult to recall an order after it had been placed. Memorizing a completed transaction was no longer necessary so they were set aside for new orders. Zeigarnik proved that a partially completed task is often more remembered than a complete task. There are two types. Short-term memory is one, and long-term memories the other. Short-term memories store information for only a short time with limited capacity. Long term memory can store unlimited amounts of data over a long period of indeterminate time. The information required for immediate use is stored in short-term memory. In the meantime, long-term information is needed for future uses. The dual-store theory of memory is well understood by most psychologists. However, there is much disagreement regarding the role of short-term or long-term memory, how they relate, and whether these two systems are truly separate.